SR-A
PSEUDO
SCIENCE

SR-A
PSEUDO
SCIENCE

Einstein's Special Theory of Relativity

Si-Xian Liang

Library of Congress Control Number:		2013901480
ISBN:	Hardcover	978-1-4797-8497-4
	Softcover	978-1-4797-8496-7
	Ebook	978-1-4797-8498-1

To order additional copies of this book, contact:
Xlibris Corporation
1-800-618-969
www.xlibris.com.au
Orders@xlibris.com.au
503105

CONTENTS

INTRODUCTION

SR is Einstein's 'Special Theory of Relativity' for short. The theory was built on the postulate of 'absolute velocity of light' that 'a light propagates simultaneously with velocity $c = 3 \times 10^8$m/s, relatively to all of inertial systems'. Here 'a light' means the identical light. The value of c is the magnitude of the emitting velocity of light relative to the light source. Obviously the postulate runs counter to classical mechanics. We will refer to the postulate as the 'Postulate' throughout this text for convenience.

Deduced under the Postulate, the *Lorentz Transformation* (LT for short) was bound to conflict with the existing *Galilean Transformation* (GT for short) and violating the primary concepts of space and time. But Einstein used 'thought experiment' to show 'the relativity of simultaneity' and hence alleged that space and time is both 'relative'. As such, it seems that the conflict could be avoidable and LT could gain 'real meaning'. He then gave LT a 'revolutionary' explanation in a unique way that would lead to SR's famous assertion of 'moving body contracts, moving clock slows' (We will refer to the assertion as the 'Assertion' throughout this text for convenience). Further development in dynamics is the so-called 'relativistic mechanics'. All of these formed Einstein's SR. As the Postulate is the pillar of SR, LT should be regarded as SR's Bible.

Over a century has passed since Einstein published the first SR paper in 1905 (Reference [1]). So far, there are still no clear reasons in theory or evidences in practice to make a convincing judgement on SR. The velocity of light is so high that it is almost impossible to conduct a direct test. 'Believe or not'? People just simply pick an option. However, time somehow has helped SR to gain broad acceptance with glory increasingly. A new and strange world displayed by SR attracted everyone's curiosity. The 'Twin Paradox' made people so excited that they ardently argued on which twin should be younger and totally forgot that paradox just is paradox. Today, SR has become a scientific truth occupying a chapter of high school syllabus of physics.

However, in eight years' exploration, the writer of this text has gradually recognised that SR actually is wrong. 'Are you serious?' Yes. This text will persuade you with powerful and convincing reasons.

In the first several years, the writer tried following Einstein's ideas but discovered that he was like being trapped in a maze of SR where nothing was certain, concepts were mixed or shifted, and logic was broken or in conflict. He eventually realised that, if you follow Einstein's path blindly, you would just keep turning rounds in the maze of SR; and you may become an expert of SR but you would never find a way out to the truth. Then he tried to think 'outside the box' and found the missing light source Galilean system which was excluded from LT by SR but was the real primary player in LT. This was a vital breakthrough, because only the light source system can be a stepping stone for the exploration of the mystic Lorentz system and hence enable us to have an insight into LT and the entire SR.

By uncovering the defects of LT, such as the dependency on light source system, and by revealing the conceptual error of 'applying LT wrongly on second party body' in the Assertion and the relativistic mechanics, this text demonstrates that SR is untenable in theory. Particularly, a shocking finding is that Lorentz time contains a local-time part which actually is a false time of no flux. Without the local-time as a fundamental stone, the whole building of SR would collapse immediately.

By rectifying the interpretation of the M-M Experiment and, particularly, by carefully analysing the Doppler Effect, this text shows a startling finding that the Doppler Effect and the Postulate are bound to be mutually exclusionary in nature, and hence verifies that SR's Postulate has been actually denied by the existing evidential practice in astronomy.

With so many fatal problems, SR has to be judged as a pseudoscience.

1. THE NATURE OF SPACE AND TIME HAS NO ROOM LEFT FOR LT

Our judgement should be based on the nature of space and time. We need first to clarify their primary concepts providing a solid foundation for the assessment of SR.

What is the primary concept of space? In our universe, all material bodies (including all celestial bodies and their attachments) relatively float, suspend, and move in a non-material common space that is like an empty, infinite room with no boundary and centre in three dimensions of isotropy and homogeneity. The universe consists of all material bodies and the common space (including all non-material energy there). Apart from the universe, there is nothing else existing (no outside of the universe). Everything has a limit on its size (space) and its lifespan (time) except the universe.

It is impossible to measure the absolute position of anybody in the common space because the common space is unable to be attached with a spatial 3D coordinate system due to its non-material emptiness. Therefore, anybody's position (and hence motion) relative to the common space could never be determined, i.e. no absolute position and hence no absolute motion (or rest). So, the common space is not a mechanics space.

However, everybody can be attached with a spatial 3D coordinate system as a reference frame in which the position of all the others can be described relatively to that body of reference. Thus, there is relative position and, hence, relative motion (or rest) in the universe only. So, the attached coordinate system representing the body's external space is a mechanics space with relativity. The attached system of a body is open to all the others in motion relative to that body which plays a role of a reference.

A body itself is unable to move into its own attached space, as both are always stuck together in a relatively stationary status (just like you cannot carry yourself on your back). The relative position between a body and its attached 3D coordinate system could be chosen freely as long as there is no relative motion between them. That means the origin of the attached coordinate system is unnecessary to correspond to the position of the body.

Bodies that are stationary relatively to each other may be commonly attached with one coordinate system for simplicity, as their positions with respect to their common coordinate system do not matter. Hence, they together will be regarded as one body carrying one spatial coordinate system.

The relativity of motion is based on the concept of attached space. For two bodies A and B in relative motion, Body A is moving in Body B's attached space that is the reference system for the motion of Body A, while Body B is moving in Body A's attached space that is the reference system for the motion of Body B. As such, the two bodies could not encounter each other in either the attached space of theirs. If the two bodies collide, the collision occurs in neither their attached space but should be considered as happening in another body's attached space, which is regarded as the two bodies' common reference system.

In general, a reference body does not encounter and interact with the others in its own attached space. Bodies can encounter and interact in their common reference system only. This is a very important concept about attached space. Violating this concept is the error we often see in SR when applying LT (see Sections 4 and 7).

Usually, a body's attached system is hidden. Only when there is a need for the body to be taken as a reference, then its attached system has to be opened and itself can be hidden. For example, the two colliding bodies have no need to open their attached system; they are colliding with the velocities relative to their common reference system of another body.

Everybody's attached space completely overlaps with the common space, and hence, all bodies' attached spaces also overlap together. The common space is all bodies' overlapping space. As bodies move relatively, the attached spaces they carry follow to move too; but they still keep overlapping together with the common space.

The concept of overlapping attached spaces is the key point to understand the structure of space. It means that, bodies' attached spaces are all overlapping together dynamically so that everybody is moving in each of the overlapping attached spaces of all other bodies simultaneously with different relative velocities.

All 3D coordinate systems should have the inherent specificity of the common space: isotropy and homogeneity. Each of the 3D coordinates should be absolutely independent of one another. Such 3D coordinate system may be specified as 'Galilean space' for contrasting with 'Lorentz space', where the coordinate system is apparently dilated in one direction (see Sections 3 and 4).

Notice that the attached space of a body also is the body's gravitational field. It may be an electrical field or a magnetic field too, if the body is an electricity charge or a magnet respectively. Since a moving body carries its attached space, its fields would follow to move together. For electromagnetic wave emitted from light source, it is a composite motion. On one hand, the wave is propagating with c, relatively to the light source, in the attached system of the light source; on the other hand, the attached system of the light source containing the electromagnetic field formed by the wave is moving together with the light source in the attached space of another body with velocity relative to that body. This concept of overlapping space is violated by SR's Postulate. In section 7 we will show the latter is wrong.

What is time? Simply to say time is a sign of change, as all changes occurring in real things (such as bodies and events in physical phenomena, chemical reactions, life developments, and human activities) are always accompanied with time elapsing. We may say that, no change, no concept of time.

The universe with infinite space consists of infinite material things and is always changing as long as at least one thing is changing. That means time elapses continuously, one instant by one instant, as the universe changes constantly bit by bit.

What does an 'instant' mean? At an instant, all things stay put and the whole universe is unchanging ('frozen') temporarily. Hence, when the universe is at the present instant, everything currently is at the present instant, and when the universe is at any instant, everything then must be at the same instant. This means 'instant' can be regarded as 'universe instant', representing the simultaneity in the whole universe for everything and everywhere. Especially all bodies in the universe, no matter where they are and how they are relatively moving, could always be at the identical universe instant simultaneously. We may simply refer to such simultaneousness as: 'the identity of universe instant', which gives no grant for the so-called 'relativity of simultaneity' in SR (see Section 5).

The universe instant runs homogeneously in one way from the infinity of the past to the infinity of the future, with its own course of flux that is natural and unchangeable to humans. Imitating the spatial coordinate axis, we may image a temporal axis for time flux.

The pace of the time flux could be recorded by a clock with a stable and periodic change. Because of the identity of universe instant, only one standard clock is needed for the whole universe and everything. Once time flux has been represented by the running of a standard clock, time gains a role of reference to make its mark on universe time-axis and to compare the duration for all things.

Of course, the universe time of a body should be passed on to its attached space such that every point in an attached Galilean space has the same universe time of the body. Such uniform systemic time in Galilean space may be classified as a sort of 'Galilean time' (for distinction from 'Lorentz time' which will be introduced in Section 3). Then, Galilean space and Galilean time together form a normal 4D 'Galilean system' (for distinction from 'Lorentz system' which will be introduced in Section 3). Obviously, Galilean time and Galilean space are independent of each other. But this common sense was ignored by SR in which Lorentz time was depended upon Lorentz space (We will tell you the truth in Section 5).

Due to the identity of universe instant, the duration between two instants must be equal for all things, no matter in what state of motion they are. Thus, Galilean time interval between two instants in attached Galilean space, Δt, must be the same for all bodies: A, B, C, . . . :

$$[\Delta t]_A = [\Delta t]_B = [\Delta t]_C = \ldots$$

This is referred to as 'the invariance of universe time interval' for all Galilean systems.

At an instant, an identical spatial spot in the common overlapping space will have different names corresponding to individual attached space simultaneously. For example, spatial spot P in the common overlapping space will be marked as $[P]_A$ in system A or $[P]_B$ in system B or $[P]_C$ in system C, and so on. Distance between two spatial spots P and Q in the common overlapping space will be marked as $[PQ]_A$ in system A or $[PQ]_B$ in system B or $[PQ]_C$ in system C, and so on. All these different names are for an identical distance between P and Q in the common overlapping space at an instant. So

$$[PQ]_A = [PQ]_B = [PQ]_C = \ldots$$

This is referred to as 'the invariance of overlapping distance (at an instant)' for all Galilean systems.

In a body's attached Galilean system, all others' relative velocity is regarded as a sort of 'Galilean velocity' that is defined as the rate of change of displacement over the lapse of systemic time:

Galilean velocity = the change of relative
position/the lapse of systemic time

When describing motion, both space and time are taken as reference coordinates which should not be altered by the motion; otherwise, things would go out of form. But that is just what SR had done (see Sections 3 and 4).

Now, based on the two invariance (invariance of overlapping distance and invariance of universe time-interval), a transformation of coordinates between two Galilean systems A $[t, x, y, z]$ and B $[\tau, \xi, \eta, \zeta]$, which are attached respectively to Body A and B in rectilinear translational motion with relative constant velocity v, can be derived (Fig.1):

$$\tau = t \qquad \xi = x - vt \qquad \eta = y \qquad \zeta = z \qquad\qquad (1)$$

This is referred to as Galilean Transformation ('GT' for short). For simplicity, the deduction of (1) has been conducted under a *standard configuration* that, at zero initial instant $t_0 = \tau_0 = 0$, the two systems are coincided completely; and that the v is stipulated to be always the velocity of the moving system B [τ, ξ, η, ζ] relative to the stationary system A [t, x, y, z] and is pointing at the positive direction of the coincided x-ξ axis.

The physical significance of GT is that it describes the dynamic relation of space and time between the two bodies' attached systems in the relative motion; but only third party bodies can enjoy the relation.

In this text we regard the stationary body as *the first party* and the moving body as *the second party*; a body moving with respect to both the first and the second party is referred to as *a third party body*. In Section 2, we will see LT relies on an initial light flash as a third party.

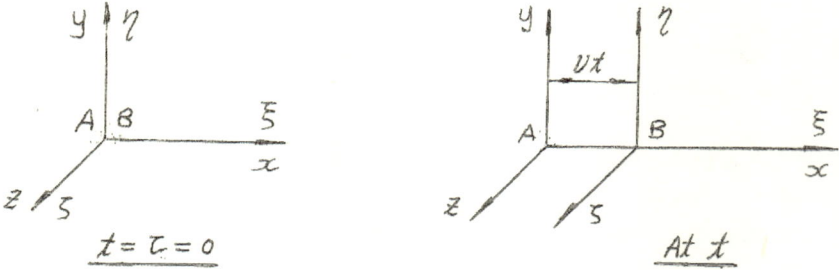

Fig.1 Galilean Transformation

One can see that GT is the direct consequence from the primary concepts of time and space. It is as natural as coordinate shift between two systems and as simple as no third party body is needed to be involved in the deduction.

Note that GT is conducted for an identical spatial spot of the overlapping space between [x, y, z] and [ξ, η, ζ] at an identical instant $\tau = t$. This is a tacit principle and may be referred to as the *Transformation Principle* of 'identical spot at identical instant'. Under the Transformation Principle, GT is the sole possible result complying with the two of invariance (invariance of overlapping distance and invariance of universe time interval). That means there is no room left for any others, such as LT, to exist, yet SR alleged that LT could supersede GT.

Unless things involve dynamics issues, the two systems are not necessarily inertial, because GT is just kinematic transformation.

Since the two bodies have a role of reference, they may be hidden and left their attached system engaging in GT for the motion of third party bodies.

We must bear in mind that *kinematic transformation (GT or LT) is for third party only*; neither first party nor second party can be allowed to take the equations of transformation. In other words, only third party can possess the space and time of the two systems simultaneously. This is another very important concept about overlapping attached space in kinematic transformation. We will see a conceptual error dominating the entire SR. That is, confusing second party body with third party and applying kinematic transformation (LT) on the second party. In result, the second party wrongly takes the space and time of the two systems simultaneously (see Sections 4 and 7).

Now, for a third party Body P in motion relative to both systems, the velocity transformation formula would be easily deduced from GT(1):

$$w_\xi = u_x - v \qquad w_\eta = u_y \qquad w_\zeta = u_z$$

These three components can be combined to form a vector relation of velocity transformation:

$$\boldsymbol{w} = \boldsymbol{u} - \boldsymbol{v} \qquad\qquad\qquad (1a)$$

where $\boldsymbol{v}$ is the velocity vector of the moving System B relative to the stationary System A; $\boldsymbol{u}$ and $\boldsymbol{w}$ is the velocity vector of third party Body P, relative to the stationary System A and the moving System B, respectively.

Note that, in kinematic transformation, the $\boldsymbol{u}$ of third party Body P and the $\boldsymbol{v}$ of second party Body B are both lying in the stationary System A so that they can interact with each other while the $\boldsymbol{w}$ of third party Body P is lying in the moving System B and hence is unable to encounter the $\boldsymbol{v}$. Therefore in some situation the kinematic transformation is useless. For instance, when Body P collides with Body B, it is $\boldsymbol{u}$, not $\boldsymbol{w}$, to impact with $\boldsymbol{v}$ in the common reference System A. Hence the collision issues need not employ GT.

Using the concept of composite motion, instead of GT, we can also obtain the same result of (1a) under classical mechanics. The motion of each of the three bodies: Body A, Body B and Body P, could be considered as composite motion. For instance, Body P's motion with u relative to Body A can be regarded as a composite motion consisting of Body P's motion with w relative to Body B and Body B's motion with v relative to Body A, i.e. $u = w + v$ which is the same as (1a).

But we should clarify that they have distinct standpoint. In the situation when investigating the change of the frequency and wavelength of a wave with respect to a moving receiver (as the second party), since the kinematic transformation is useless for second party body we have to consider the concept of composite motion instead. In Section 7, we will see such situation in investigation of Doppler Effect in detail.

Notice that whichever of the two systems is chosen as the stationary system, GT (1) and (1a) would maintain unchanged, provided the v representing the velocity of the moving system relative to the stationary system and pointing to the positive direction of the x-ξ axis. This nature is in line with the relativity of motion. If we only swap the relative status between the two systems while retaining the coordinate systems unchanged, we would need to change the sign of v, since the direction of relative motion has been switched.

Further, consider that a third party Body P of mass m is in accelerating motion under the effect of a net force F_A in System A which is supposed to be an inertial system. How would it be in System B correspondingly?

What is an inertial system? If the net force of all external forces exerted on a body equals zero, the body is an inertial body; and its attached system is an inertial system. (This is just an idealised model in reality. The ground of Earth may be considered as an inertial system approximately). As Newton's laws of motion only hold in inertial systems, you may say that inertial system is the one in which Newton's laws can hold.

For two bodies A and B in relative, translational, rectilinear motion with constant velocity, if one of them, say A, is an inertial body, then B, the other, must also be an inertial body, because B has no acceleration relative to A and hence must have zero net force asserted on it.

Since the two systems are inertial, we can apply Newton's laws of motion on a third party Body P. Then Body P would have $F_A = ma_A$ where $a_A = \frac{d}{dt}(u)$, that is the acceleration of Body P in inertial System A, while in inertial System B, the force exerted on Body P correspondingly is $F_B = ma_B$, where $a_B = \frac{d}{d\tau}(w)$, that is the acceleration of Body P in System B (note that acceleration may have direction different to the direction of velocity vector, should it be curved motion).

As a sort of mechanical effect between bodies, force is absolutely independent of the choice of reference system. Thus, the net force exerted on Body P should be identical for the two systems: $F_B = F_A$. Then it would arrive at $a_B = a_A$. That means P should have equal acceleration in the two systems. A perfect example is that for a ball tossing motion along a vertical line on a train of constant velocity; it would become a projectile curved motion seen from the ground. The same object experiences identical gravitational force and identical acceleration in both inertial systems: the train and the ground.

On the other hand, we can easily verify that GT is compatible with Newton's laws. By differentiating the (1a) of GT, we immediately obtain: $\frac{d}{d\tau}(w) = \frac{d}{d\tau}(u) = \frac{d}{dt}(u)$, i.e. $a_B = a_A$.

Consequently, GT would satisfy the requirement that the Law of Conservation of Momentum must be valid in the two inertial systems.

Therefore, GT agrees with 'the principle about inertial systems' that the physical laws (dynamics) hold for all inertial systems (Einstein named it 'the Principle of Relativity', which would be easily mixed with others of relativity. So we re-name it focusing on inertial systems).

As we have stated, the primary concept of time widely involves in all sorts of change, not just in motion. Once the time rate had been determined by the standard clock, the change of a real thing, including the relative motion between bodies, has no influence on the flux of time. Therefore, time should be always treated as an absolutely independent reference in rating the change of a real thing. In the theory of SR, time is dependent on location and motion. That actually is just abuse on relativity. In description of motion, altering time reference, as well as space reference, is too much beyond the relativity of motion and would be unavoidable to make the picture of motion ridiculous.

However, SR selling its theory is under the banner of 'revolutionary ideas' that common sense is useless to argue with. But the wrong is wrong. Violating the primary concepts of time and space is impossible to be correct, and hence, SR is bound to contain defects and problems. The truth has to, and certainly can, be revealed.

2. LT TOTALLY RELIES ON THE MISSING LIGHT SOURCE SYSTEM

In the 1905 SR Paper [1], Einstein conducted a physical deduction for LT under the Postulate of 'absolute velocity of light'. It was a pity that his trial could not be recognised as a success. In fact, he did not investigate the three components of a light ray of an arbitrary direction. Instead, he chose three light rays along three coordinate directions to work on. Of course, special cases cannot deliver a general result. Also, he handled the time in the two systems both as uniform systemic time. Certainly, that was a mission impossible to deduce LT from two Galilean systems.

For theoretical perfection, the deduction of LT should be conducted on a ray in a general direction.

Suppose two bodies B_1 and B_2 are in rectilinear translational relative motion; their attaching 4D systems are $B_1 [\tau_1, \xi_1, \eta_1, \zeta_1]$ and $B_2 [\tau_2, \xi_2, \eta_2, \zeta_2]$ respectively (Fig.2).

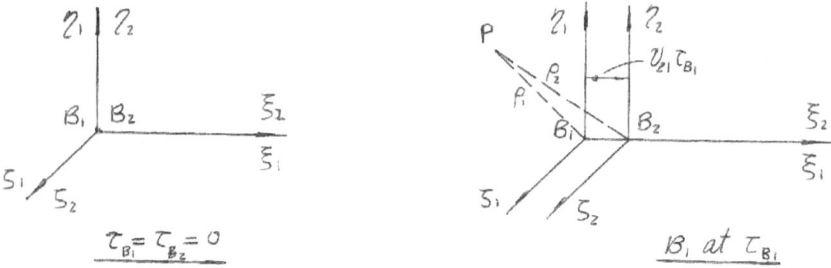

Fig.2 Lorentz Transformation (light source system is absent)

For convenience, make the two systems under *the standard configuration for LT* such that the two systems are completely coincided at zero initial instant when $\tau_{B1} = \tau_{B2} = 0$ at the two coincided origins B_1 and B_2; the v_{21} as the velocity of the moving system B_2 relative to the stationary system B_1, points to the positive direction of the coincided ξ_1-ξ_2 axis. Besides, a light flash is employed as an initial third party and emitted from the location of the coincided origins at the zero initial instant.

When a wavefront of the initial flash along an arbitrary direction arrives at spatial point P: as P $[\tau_1, \xi_1, \eta_1, \zeta_1]$ in System B_1 or as P $[\tau_2, \xi_2, \eta_2, \zeta_2]$ in System B_2, the displacement $\rho_1 = B_1P$ in System B_1 or $\rho_2 = B_2P$ in System B_2. We can have

$$\rho_1^2 = \xi_1^2 + \eta_1^2 + \zeta_1^2$$

$$\rho_2^2 = \xi_2^2 + \eta_2^2 + \zeta_2^2$$

Under Einstein's Postulate of 'absolute velocity of light':

$$\rho_1 = c\,\tau_1 \qquad \rho_2 = c\,\tau_2$$

Hence:

$$\xi_1^2 + \eta_1^2 + \zeta_1^2 = (c\,\tau_1)^2 \tag{2}$$

$$\xi_2^2 + \eta_2^2 + \zeta_2^2 = (c\,\tau_2)^2 \tag{3}$$

All rays of the flash in every radial direction together would form a wavefront sphere. Thus, (2) or (3) not only shows the square of displacement of the flash along one radial direction from origin B_1 or B_2 reaching the point P, but also represents the equation of the entire wavefront sphere in System B_1 or B_2, respectively. The sphere must be identical between the two systems since the point P is identical according to the Transformation Principle of 'identical spatial spot at identical instant'.

A problem emerges: Two equations simultaneously representing an identical sphere would generate contradiction in space and time between the two systems. To resolve the problem, an anti-GT priori presumption is needed as below:

$$\tau_2 = m\,\tau_1 + n\,\xi_1 \qquad \xi_2 = \beta_{21}\,(\xi_1 - v_{21}\,\tau_1) \qquad \eta_2 = \eta_1 \qquad \zeta_2 = \zeta_1$$

where v_{21} is stipulated to be the velocity of the moving system (B_2) relative to the stationary system (B_1) while m, n, and β_{21} are all constants to be found.

Substituting the presumption into (3), then equating the coefficients in comparison with the (2), the three constants m, n, and β_{21} can be determined. As a result, a set of four equations is derived:

$$\tau_2 = \beta_{21}\,(\tau_1 - \xi_1 v_{21}/c^2) \quad \xi_2 = \beta_{21}\,(\xi_1 - v_{21}\,\tau_1) \quad \eta_2 = \eta_1 \quad \zeta_2 = \zeta_1 \qquad (4)$$

where $\beta_{21} = c/\sqrt{c^2 - v_{21}^{\,2}}$, named as 'Lorentz Factor'. Notice that $v_{21} < c$ and $\beta_{21} > 1$.

The set (4) is the result brought up by the transforming of the motion of a light flash between two systems under the Postulate. It obviously conflicts with GT. But SR believes that a 'new continent' has been discovered such that the result can be extended to be a general transformation law for any third party body, and hence refers to it as Lorentz Transformation (LT).

In the LT (4), we only roughly know the two systems are so strange that time τ varies with different coordinate ξ. Both τ_1 and τ_2 is not a sort of Galilean time and hence cannot link with universe time. That means both systems are non-normal 4D space. Distinguishing with the Galilean system, they may be referred to as 'Lorentz system' [τ, ξ, η, ζ] with 'Lorentz space' (ξ, η, ζ) and 'Lorentz time' τ.

Consequently, all velocities relative to a Lorentz system would belong to a sort of 'Lorentz velocity' that is distinct to Galilean velocity. As time differs from point to point in Lorentz system, a more general definition of velocity is needed to suit Lorentz velocity but not to conflict with Galilean velocity.

Generally, if a body is moving from point P at time τ_P to point Q at time τ_Q in a system, then the velocity of the body along path PQ can be defined as below:

Velocity = (The spatial interval of PQ)/The time interval ($\tau_Q - \tau_P$)

This general definition can cover Lorentz velocity as well as Galilean velocity.

Since both B_1 and B_2 are Lorentz system, the relative velocity v_{21} (or v_{12}) between the two systems would be Lorentz velocity satisfying the above definition. For calculation of v_{21}, the spatial interval of B_2B_1 in system B_1, representing the displacement of B_2 relative to system B_1, equals ξ_1 of origin B_2; the corresponding time interval equals τ_1 of origin B_2. So $v_{21} = (\xi_1/\tau_1)$ where ξ_1 and τ_1 are the coordinates of origin B_2 $[\tau_1, \xi_1, 0, 0]$ in System B_1.

Differentiating the first equation of (4), we get:

$$\frac{d\tau_2}{d\tau_1} = \beta(1 - u_x v_{21}/c^2) = \beta N \qquad (5)$$

where $N = 1 - u_x v_{21}/c^2$.

Then differentiating the other three equations of (4), we obtain third party's velocity transformation formula as:

$$w_\xi = (u_x - v_{21})/N \qquad w_\eta = u_y/\beta N \qquad w_\zeta = u_z/\beta N \qquad (6)$$

where $u(u_x, u_y, u_z)$ and $w(w_\xi, w_\eta, w_\zeta)$ is the velocity of a third party relative to System A and System B, respectively.

If the velocity of a third party (anything), relative to system A in whatever direction, is given as $u = c$, then by substituting it into (6), the velocity of the third party relative to system B would be $w = c$ too. This result of $w = c = u$ is no surprise, because LT was derived under the Postulate of 'absolute velocity of light'. With such a nature, LT should be treated as the Bible for SR in theory.

Note that the (6) is unable to be converted into vector expression because there is a contradiction between the magnitude and the direction of **w**, which will be demonstrated in Section 3.

An interesting question arises: Where is the wavefront sphere? It was so mystic that no illustration of the wavefront sphere could be found in the articles about LT. What a strange phenomenon!

To answer the question, we must first find the light source that emits the initial flash.

Since the flash must be emitted from a light source, there should be a *light source system* A [*t, x, y, z*] that had been missing or neglected by SR. The two Lorentz systems B_1 [$T_1, \xi_1, \eta_1, \zeta_1$] and B_2 [$T_2, \xi_2, \eta_2, \zeta_2$] must be both in parallel motion relative to the light source System A.

When the light source emits the initial flash at the origin in System A, all rays of the flash in every radial direction must be equivalent and point-symmetric over origin A. Therefore, no doubt the light source System A is a primary Galilean system whose uniform systemic time *t* links with universe instant.

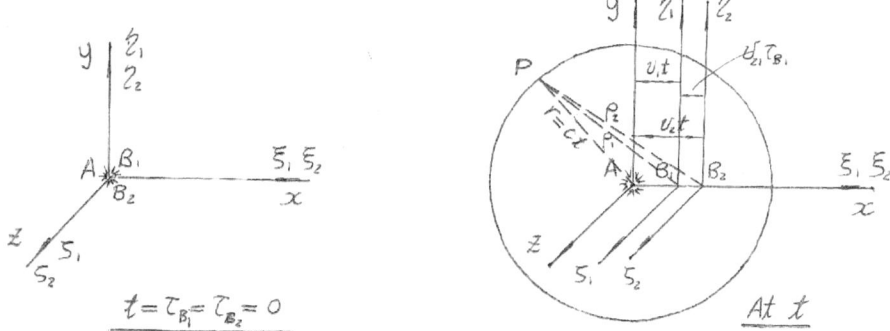

Fig.3 Lorentz Transformation (hosted by light source system A)

Under the standard configuration (Fig.3), the light source is located at the origin in Galilean System A; the three systems, A, B_1, and B_2 completely coincide at zero initial instant $t = T_{B1} = T_{B2} = 0$, the time when the initial flash is emitted from the light source.

While the equation (2) and (3) of the wavefront sphere retain unchanged for System B_1 and B_2 respectively, the wavefront sphere in System A at any time *t* has equation:

$$x^2 + y^2 + z^2 = (c\,t)^2 \qquad\qquad (7)$$

where $(c\,t) = r\ (= AP)$. As Galilean time *t* is unified for the whole System A, *r* would be consistent in all radial directions and hence is the true geometric radius of the sphere with centre at origin A. That means the wavefront sphere has a normal mathematic sphere-equation (7) in the light source system so that it can be drawn as in Fig.3.

There may be a series of Lorentz systems, including $B_1 [\tau_1, \xi_1, \eta_1, \zeta_1]$ and $B_2 [\tau_2, \xi_2, \eta_2, \zeta_2]$, leading by a light source system A $[t, x, y, z]$. All of them are in uniform rectilinear motion relative to and parallel to each other; each of them has different velocity relative to A. With the standard configuration, the light source emits the flash from the coincided origins at the initial instant $t = \tau_{B1} = \tau_{B2} = \ldots = 0$. Then, the initial light propagates simultaneously in the overlapping attached spaces including Galilean System A and all Lorentz systems of the 'B' series (see Section 1).

Under the Postulate of 'absolute velocity of light', the LT between light source Galilean System A $[t, x, y, z]$ and any one of Lorentz System B $[\tau, \xi, \eta, \zeta]$ of the 'B' series can be deduced in similar way or simply by rewriting the LT(4), as below (Fig.4):

$$\tau = \beta \left(t - \frac{v}{c^2} x\right) \qquad \xi = \beta (x - v t) \qquad \eta = y \qquad \zeta = z \qquad (8)$$

where Lorentz Factor $\beta = \frac{c}{\sqrt{c^2 - v^2}}$. The v is stipulated as the velocity of the particular System B relative to the stationary light source System A and should be a Galilean velocity (because second party body B is moving in Galilean System A).

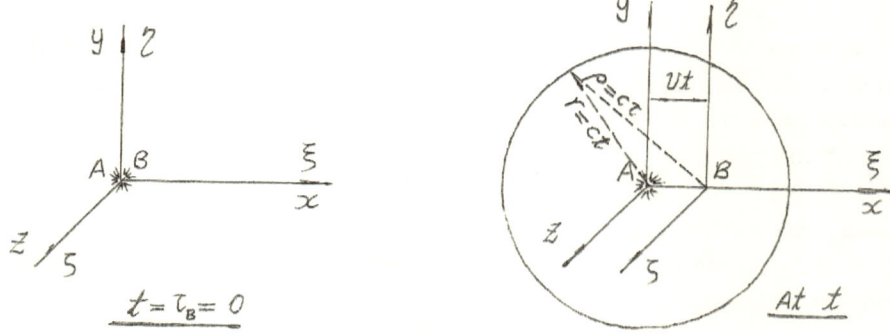

Fig.4 Lorentz Transformation (light source system is involving)

Under LT(8), the third party's velocity transformation formulae would have the same form as the (6):

$$w_\xi = (u_x - v)/N \qquad w_\eta = u_y/\beta N \qquad w_\zeta = u_z/\beta N \qquad (8a)$$

but with $N = 1 - u_x v/c^2$.

If the status of relative motion is swapped between the two systems but maintain the configuration unaltered, the forms of LT(8) and (8a) will retain the same, provided the sign of v changes. This verifies LT complies with the relativity of motion.

Since the wavefront sphere emerges in the related light source Galilean system as described mathematically by equation (7), now the mystery can be cleared up.

According to the Transformation Principle of 'identical spot at identical instant', the equations (2), (3) and (7) must describe the same wavefront (Fig.3). Thus there is only one identical wavefront sphere expressed mathematically by the (7), appearing in the sky of overlapping space, although each system of the 'B' set has a different interpretation about it.

Each system of the 'B' set with different relative velocity v and Lorentz factor β would claim that the wavefront sphere is emitted at the initial instant from its own origin B physically, but at t, the geometric centre of the sphere is at $\xi = -\beta vt$ with geometric radius $r = ct$. That is, for each system of the 'B' set, the mathematical equation of the sphere is derived by substituting the second, the third and the fourth equations of LT(8) into the (7):

$$(\xi + \beta vt)^2 + \eta^2 + \zeta^2 = (ct)^2 = constant$$

While by substituting the first equation of LT(8) into the (2) or (3), the result may be regarded as the physical equation of the wavefront with physical centre at origin B (More precisely, it is the point of System B that had been coincided with origin A at the initial instant when the flash was emitted) and variable 'physical radius' ρ:

$$\xi^2 + \eta^2 + \zeta^2 = (c\tau)^2 = c^2[(t/\beta) - (v/c^2)\xi]^2 \qquad (9)$$

where $\rho = c\tau = c[(t/\beta) - (v/c^2)\xi] \neq constant$.

As τ is a function of ξ, the 'physical radius' p is variable in System B. That is why the flash sphere cannot be drawn in System B when the light source System A is absent.

We may say that the flash sphere in the Lorentz system B is 'physically eccentric' with eccentric degree $e = vt/r = v/c = constant$. The eccentric degree would maintain unchanged as time passes (Fig.4).

Now let's check if the LT(8) between light source Galilean system and a Lorentz system could be compatible with the LT(4) between two Lorentz systems.

Under the standard configuration (Fig.3), let v_1 and v_2 respectively be the velocity of Lorentz Systems B_1 and B_2, relative to the related light source System A, both pointing to the positive of x-ξ_1-ξ_2 axis. They must be Galilean velocity since A is a Galilean system.

Apply LT (8) between A and B_2:

$$\tau_2 = \beta_2\,(t - v_2\,x/c^2) \quad \xi_2 = \beta_2\,(x - v_2\,t) \quad \eta_2 = y \quad \zeta_2 = z \quad (10)$$

where $\beta_2 = c/\sqrt{c^2 - v_2^{\,2}}$

Also apply LT (8) between A and B_1:

$$\tau_1 = \beta_1\,(t - v_1\,x/c^2) \qquad \xi_1 = \beta_1\,(x - v_1\,t) \qquad \eta_1 = y \qquad \zeta_1 = z$$

i.e.

$$t = \beta_1\,(\tau_1 + v_1\,\xi_1\,/c^2) \quad x = \beta_1\,(\xi_1 + v_1\,\tau_1) \quad y = \eta_1 \quad z = \zeta_1 \quad (11)$$

where $\beta_1 = c/\sqrt{c^2 - v_1^{\,2}}$

Then the transformation between B_1 and B_2 can be derived by substituting (11) into (10):

$$\tau_2 = \beta_1\,\beta_2\,(1 - v_1\,v_2\,/c^2)[\tau_1 - (\frac{v_2 - v_1}{1 - v_1 v_2\,/c^2})\,\xi_1\,/c^2] = \beta_{21}\,(\tau_1 - v_{21}\,\xi_1\,/c^2)$$

$$\xi_2 = \beta_1\,\beta_2\,(1 - v_1\,v_2\,/c^2)[\xi_1 - (\frac{v_2 - v_1}{1 - v_1 v_2\,/c^2})\,\tau_1] = \beta_{21}\,(\xi_1 - v_{21}\,\tau_1)$$

$$\eta_2 = \eta_1$$

$$\zeta_2 = \zeta_1$$

where

$$v_{21} = (v_2 - v_1) / (1 - v_1 v_2 / c^2) \qquad (12)$$

$$\beta_{21} = \beta_1 \beta_2 (1 - v_1 v_2 / c^2) = c / \sqrt{c^2 - v_{21}^2} \qquad (13)$$

In comparison between the (4) and the above result, we can see that β_{21} of the (13) is the same as the Lorentz factor in the (4); and we can also make a judgement that v_{21} of the (12) must be the same Lorentz velocity of B_2, relative to B_1, as in the (4), because System B_2 can be regarded as a third party in the LT between System A and System B_1.

We may use the definition of Lorentz velocity to verify v_{21} of B_2.

For origin B_2 in Lorentz System B_1, it is at $\xi = 0$ and $\tau = 0$ initially (when $t = 0$); and it is at $\xi = \beta_1(x - v_1 t) = \beta_1(v_2 - v_1) t$ and $\tau = \beta_1(t - x v_1 / c^2) = \beta_1 t(1 - v_2 v_1 / c^2)$ as it has moved a displacement $x = v_2 t$ in Galilean System A at universe instant t. So we can obtain Lorentz velocity of B_2, relative to B_1, as $\Delta\xi/\Delta\tau = (v_2 - v_1)/(1 - v_1 v_2 / c^2) = v_{21}$ indeed.

These results have verified that the LT (8) between light source Galilean System A and a Lorentz System B has no conflict with the LT (4) between two Lorentz systems; v_{21} is the Lorentz velocity of B_2 moving in the attached space of B_1 relatively to B_1 and must be defined by the (12). Notice that v_{21} is distinct to $(v_2 - v_1)$ which is the Galilean velocity difference between B_2 and B_1 moving in the Galilean System A.

It should be pointed out that SR had never mentioned the light source Galilean system involving in LT, but SR did actually treat one of the two systems as a Galilean system. This is because without the Galilean system as a stepping stone, no one can find out what a Lorentz system would be concretely.

LT as a kinematic transformation shows only the relation between two systems but no information about individual systems in detail. Look at LT

(4) between two Lorentz Systems B_1 and B_2. Each of the four quantities τ_2, ξ_2, τ_1, and ξ_1 is a function of another two, like interlocking rings could not be resolved.

While for LT (8) between a Galilean System A and a Lorentz System B, t is linked with universe time and is independent of x. At a certain time t, both τ and ξ can be determined by an arbitrarily given x. Thus Lorentz system can be ascertained by means of Galilean system (we shall discuss this in detail later).

In fact, SR's Assertion of 'moving rod shortens, moving clock slows' and the relativity of simultaneity are all based on comparison with Galilean system which rightfully is the missing light source system. If no light source system to be a guide, the Lorentz system would remain mysterious. Therefore, the light source system must be called back and must be labelled to avoid confusion in SR.

If choose a different light source System for the pair of Lorentz Systems B_1 and B_2 it would result in different v_{21} and β_{21} and hence different LT, due to different $(v_1 v_2)$, even if $(v_2 - v_1)$ may retain the same. Therefore, the LT between any two Lorentz systems is depended upon the selection of a light source system from the series of bodies which are all in uniform rectilinear motion relative to and parallel to each other.

In general, considering anybody could be a light source system and a body of non-light source could be attached with different Lorentz systems related to different light source systems and further taking all directions of relative motion into account, we could conclude that everybody's external space could be attached with a sole Galilean system and countless distinct Lorentz systems simultaneously: every spatial spot would possess countless different times; every spatial interval would have countless different lengths (see Section 3). What an unthinkable world! It could not be reality, could it? With such dependence on light source which is freely chosen, LT has no certainty to be trusted.

Therefore, the existence of a light source system, on one hand, would help to understand the Lorentz system; on the other hand, would make LT uncertain.

Since only the LT (8) between a light source System A and a Lorentz System B could be definite (Fig.4), our further discussion shall have to be restricted on LT (8) with light source stationary System A involving, unless stated otherwise.

3. THERE IS NO WAY FOR LT TO GAIN 'REAL MEANING'

Since Galilean System A has systemic time t that links with universe time, our exploration in Lorentz space should be carried on at an arbitrary time t. Considering $\eta \equiv y$ and $\zeta \equiv z$, we will only focus on $\xi = \beta(x - vt)$, the second of LT(8).

At time t, the length between any two spatial spots in overlapping space would be

$$\Delta\xi = \beta\,\Delta x \qquad \text{(condition: } t = constant) \qquad (14)$$

Obviously, it violates the primary concept of space on 'invariance of overlapping distances (at an instant)'. Take the distance between the two origins, AB, as example (Fig.4). The distance of AB in the space of Galilean System A is $[AB]_A = |x_B - x_A| = vt$, while the distance of AB in the space of Lorentz System B is $[AB]_B = |\xi_B - \xi_A| = \beta vt$. We see at once that $[AB]_B = \beta [AB]_A$ indicating the identical distance AB has two different lengths: $[AB]_B \neq [AB]_A$. Thus, the (14) is like a 'deadlock' of LT.

To unlock the 'deadlock' means to make '$[AB]_A = vt$' and '$[AB]_B = \beta vt$' be both correct in the sense of relativity. To fulfil that, the only possible way is to let the geometric relation between $[AB]_A$ and $[AB]_B$ transmitted from coincidence in space to a manner of *projection*, that is, to project spatial points between the two spaces with the (14) as the rule of projection. As Galilean space is already known, how would Lorentz space be under projection?

According to the (14), Lorentz space will dilate, with respect to Galilean space, by a factor of (β) along ξ-axis linearly, but the relative position between the two spaces needs to be determined first. From the (13), we get $x = vt$ when $\xi = 0$. The η-ζ plane at $\xi = 0$ is referred to as *the central plane of the Lorentz space*. The projection must retain the central plane fixed on to the existing Galilean space, i.e. lap-riveted on y-z plane at $x = vt$ when it is time t. Then, each of all the other coordinate points of the Lorentz system on both sides of the central plane is projected correspondingly from the overlapping Galilean spatial spot to a new location such that the Lorentz space dilates outwardly.

Take the interval AB as an example. Its B end at $\xi = 0$ is lap-riveted on $x_B = vt$, while its A end is projected to a new location at $\xi_A = -\beta vt$, i.e. origin A would no longer coincide with its corresponding point of System B. That means System A's position in System B is projected away relatively to the origin B.

For those intervals having no end point on the central plane, their two end points are all projected to new locations.

Now we turn to investigate the time in the Lorentz system. By substituting the second into the first of LT(8), we derive a vital relation:

$$\tau = \beta \left(t - \frac{V}{c^2} x\right) = t/\beta - \frac{V}{c^2} \xi \qquad (15)$$

It clearly demonstrates that Lorentz time τ at every point ξ in Lorentz System B consists of two parts: (t/β) and $\left(-\frac{V}{c^2} \xi\right)$.

The first part (t/β) is depended on the light source Galilean System A's universe time t only and has no link with spatial location. Hence, it is systemic uniformly in the whole Lorentz System B and can be referred to as Lorentz time's 'systemic time part'. At a fixed point ξ of Lorentz space for the temporal interval Δt, we get from the (15):

$$\Delta\tau = \Delta t / \beta \qquad \text{(condition: } \xi = constant) \qquad (16)$$

That $\Delta\tau$ in the (16) represents the corresponding interval of Lorentz systemic time. Obviously, the (16) violates the primary concept of time on 'invariance of overlapping time-interval'. Thus the (16) is another 'deadlock' of LT.

Similar to the situation on space relation (14), projection is the only possible way to unlock it. As Galilean time is already known as the same as universe time, how would the Lorentz systemic time be under projection?

According to the (16), Lorentz systemic time in System B will contract, with respect to Galilean time t in System A by a factor of $(1/\beta)$ along its time-axis linearly. Considering the two systems have both taken zero initial systemic time, we must retain it as the common temporal point for the two systems while each of all other past and future temporal points of it is projected correspondingly from the overlapping Galilean time to a new temporal point such that the Lorentz systemic time contracts inwardly to it. Under projection, every Lorentz systemic time would be projected to a differing instant except it is the common zero initial instant.

The second part ($- \frac{V}{c^2} \xi$) is depended only on ξ, differing between different η-ζ planes. We refer to it as 'local-time part' of Lorentz time τ. It is the local-time part that makes the total Lorentz time τ differing between the η-ζ planes in Lorentz System B. The η-ζ plane with ($- \frac{V}{c^2} \xi$) = 0 may be referred to as *the zero local-time plane*. Obviously, the zero local-time plane is coinciding with the central plane at $\xi = 0$, where Lorentz time $\tau \equiv (t/\beta)$, containing systemic time only (under standard configuration of LT). Hence, the Lorentz time at origin B ($\xi = 0$) always represents the systemic time part for the whole Lorentz system. Outwardly, from both sides of the zero local-time plane, local-time is assigned linearly along ξ-axis with positive value at $\xi < 0$ and negative value at $\xi > 0$.

For since there is no local-time in the Galilean system, Lorentz local-time, as a specific feature of the Lorentz system, is not subject to projection (We will focus on it in Section 5).

Needless to say, with the two 'deadlocks' (14) and (16), LT itself must be a paradox with no solution; unlocking the 'deadlocks' by projection is just a temporary solution for demonstrating the strange Lorentz system conveniently. At the end of the day, we will judge the projection is impossible.

At present, the features of Lorentz system could be described as follows:

- Under projection, the Lorentz space dilates outwardly from both sides of its geometrically central η-ζ plane at $\xi = 0$, which is lap-riveted with Galilean space's y-z plane at $x = vt$, by a factor of (β) along ξ-axis linearly in comparison to Galilean space. That means in the Lorentz system, the scale of ξ-axis is enlarged β time (Fig.5a);

- Under projection, the Lorentz systemic time contracts inwardly from both sides of the common temporal point of $(t/\beta) = 0$, which is lap-riveted with Galilean time at $t = 0$, by a factor of $(1/\beta)$ along systemic time axis linearly in comparison to Galilean time. That means that in the Lorentz system, the scale of systemic time axis is reduced β time (Fig.5b);

- In addition to the systemic time (t/β), local-time ($-\frac{V}{c^2}\xi$) is another part of Lorentz time τ. Lorentz local-time ($-\frac{V}{c^2}\xi$) varies linearly along ξ-axis, differing between η-ζ planes with zero local-time plane at $\xi = 0$ under LT's standard configuration, while there is no local-time in the Galilean system at all (Fig.5c).

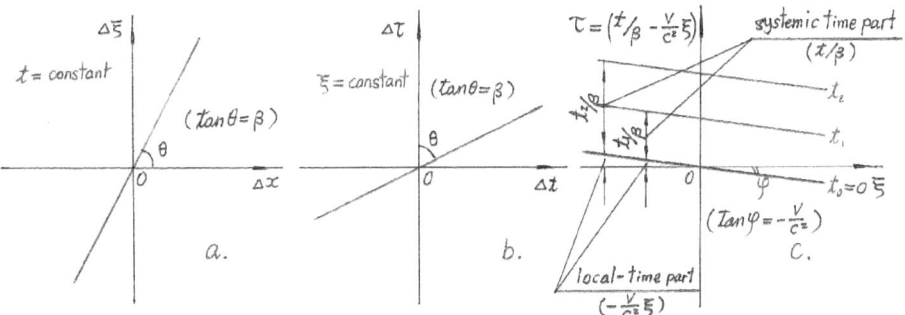

Fig.5 The features of Lorentz system

This is the result of our exploration on the Lorentz system. We have seen that, without comparison to the Galilean system, the Lorentz system could not be ascertained. Hence, LT must have the light source system presented. In fact, SR had always implicitly taken the stationary system as a Galilean system but just did not admit it being the light source system.

It should be pointed out that the concept of Lorentz local-time also has been missing in SR all the time. SR just considered Lorentz space and systemic time but had not mentioned the part of Lorentz local-time (only noticed the phenomenon about the relativity of simultaneity but did not realise the cause is the existence of the local-time). We will discuss the local time in detail in section 5.

Now the relation between the two systems, no matter which light source system it is, can be fixed in words according to the (14) and the (16) as below:

Lorentz length = β(Galilean length)
$$\text{(at a certain Galilean time)} \qquad (17a)$$

Lorentz systemic time = (Galilean time)/β
$$\text{(at a fixed spot of the Lorentz system)} \quad (17b)$$

From the (17a) and the (17b) we know that the nature of space and time of the two systems, under LT(8), is no longer dependent upon their status in the relative motion, but is determined on whether the system is a light source or not. Therefore, the existence of the light source system has destroyed the so-called 'relativity of space and time', which is one of SR's revolutionary ideas.

We have explored the Lorentz system in a manner of projection in order to open the two 'deadlocks': the (14) and the (16) of LT. But in reality, such a projection is unallowable and unachievable. Why? The reasons are stated below.

We have known that, under a projection idea, an event at point (x, y, z) and at instant t in Galilean System A is projected to the location at point (ξ, η, ζ) and to the time at instant τ in Lorentz System B; the point (x, y, z) and the point (ξ, η, ζ) do not coincide in overlapping space; the instant t and the instant τ are not the identical universe instant. Therefore, the spatial and temporal location of the event has respectively changed in the overlapping space and on the universe time line, and hence after projection, an event has actually become another different event. Such projection not only would make LT losing practical significance for the purpose of science, but also would violate the Transformation Principle of 'identical spot at identical instant'.

From the deduction of LT, we knew that, at one universe instant t, only one spherical wavefront of a flash appears in the overlapping space and is identical for all systems. But with the projection idea, every Lorentz system would have its own ellipsoidal egg-like wavefront projected from the original spherical wavefront in the light source system. Countless Lorentz systems would have countless ellipsoidal egg-like wavefronts occurring in the sky. Such ridiculous phenomena conflicts with the condition for the deduction of LT and will kill LT before its birth, because sphere equation (2) or (3) cannot hold for ellipsoidal egg-like wavefronts.

We can see that, under no circumstance, could the Transformation Principle of 'identical spot at identical instant' be abandoned, and hence, the projection idea is unallowable and unrealisable and has to be given up.

Therefore, deadlocks in LT remain as deadlocks with no possibility to be unlocked. It is clear that having these two 'deadlocks' (not to mention the specific local-time, which will be discussed in Section 5), Lorentz system cannot be real.

The deadlocks would certainly generate more defects in SR. Let's continue our exploration on LT to find other defects.

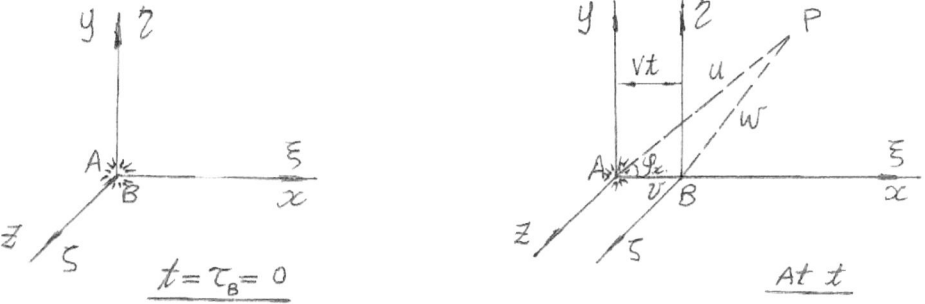

Fig.6 Third party body P in LT

Consider the motion of a third party body from origin A to point $[P]_A$ (x, y, z) in Galilean System A during time increment from 0 to t (Fig.6). Correspondingly, in Lorentz System B, which is in motion relative to System A with velocity v pointing to the positive direction of the coincided x-ξ axes, the body moves from origin B to point $[P]_B$ (ξ, η, ζ) during time increment from 0 to τ.

In the time interval from 0 to t, the displacement of origin B is $[AB]_A = vt$ in System A; the path of the body P is $[AP]_A = r = \sqrt{x^2 + y^2 + z^2}$ in system A or is $[BP]_B = \rho = \sqrt{\xi^2 + \eta^2 + \varsigma^2}$ in System B.

If for the case under GT, $[\triangle APB]_A$ in System A and $[\triangle APB]_B$ in System B are completely coincided so that we can say that the (vt), the r and the ρ together form a path triangle $\triangle APB$, which also can be regarded as a velocity vector triangle formed by the vector $\boldsymbol{v}$, the vector $\boldsymbol{u}$ of Body P relative to System A, and the vector $\boldsymbol{w}$ of Body P relative to System B since $t = \tau$. Therefore, by substituting GT(1) or by applying triangle cosine law in $\triangle APB$, we can get:

$$\rho = \sqrt{r^2 - 2rvt \cos \varphi_x + (vt)^2}$$

$$w = \sqrt{W_\xi^2 + W_\eta^2 + W_\varsigma^2} = \sqrt{u^2 - 2uv \cos \varphi_x + v^2}$$

and their three direction angles (ρ and w are in the same direction):

$$\varphi_\xi = cos^{-1}(w_\xi / w) = cos^{-1}[(u \cos \varphi_x - v) / \sqrt{u^2 - 2uv \cos \varphi_x + v^2}]$$

$$\varphi_\eta = cos^{-1}(w_\eta / w) = cos^{-1}[u \cos \varphi_y / \sqrt{u^2 - 2uv \cos \varphi_x + v^2}]$$

$$\varphi_\varsigma = cos^{-1}(w_z / w) = cos^{-1}[u \cos \varphi_z / \sqrt{u^2 - 2uv \cos \varphi_x + v^2}]$$

where φ_x, φ_y, and φ_z are the three direction angles of r (or u) over x-axis.

But for the case under LT, $[\triangle APB]_A$ in Galilean System A and $[\triangle APB]_B$ in Lorentz System B are not coincided at all. As such, neither path triangle nor velocity vector triangle can be formed (That is why the velocity transformation formulae (6) are unable to form vector expression for the whole velocity vector). Therefore we cannot apply triangle cosine law but substitute LT(8) only to get:

$$\rho = \sqrt{\xi^2 + \eta^2 + \varsigma^2} = \sqrt{\beta^2 (x - vt)^2 + y^2 + z^2}$$

$$= \beta \sqrt{r^2 - 2rvt \cos \varphi_x + (vt)^2 M}$$

$$w = \sqrt{W_\xi^2 + W_\eta^2 + W_\varsigma^2} \quad (\text{or} = \rho / \tau)$$

$$= \sqrt{u^2 - 2uv \cos \varphi_x + v^2 M} \ / \ (1 - \frac{V}{C^2} u \cos_x)$$

and their three direction angles (ρ and w are in the same direction):

$$\Phi_\xi = \cos^{-1}(w_\xi /w) = \cos^{-1}[(u \cos \varphi_x - v)/ \sqrt{u^2 - 2uv \cos \varphi_x + v^2 M}\,]$$

$$\Phi_\eta = \cos^{-1}(w_\eta /w) = \cos^{-1}[u \cos \varphi_y / \beta \sqrt{u^2 - 2uv \cos \varphi_x + v^2 M}\,]$$

$$\Phi_\zeta = \cos^{-1}(w_z /w) = \cos^{-1}[u \cos \varphi_z / \beta \sqrt{u^2 - 2uv \cos \varphi_x + v^2 M}\,]$$

where $M = 1 - \sin^2 \varphi_x u^2/c^2$.

By comparing the two cases, we can see that:

- Unlike GT, LT is incompatible to triangle cosine law, and hence, conventional coordinate geometry does not hold in Lorentz space.

- According to the Transformation Principle of 'identical spatial spot at identical instant', the three direction angles (Φ_ξ, Φ_η and Φ_ζ) under LT must be no different to those (φ_ξ, φ_η and φ_ζ) under GT. But now, the result has shown that they are distinct. That means velocity vector of w contains a contradiction between magnitude and direction. Thus u, v and w are unable to form a triangle. What a disorder has been made by LT!

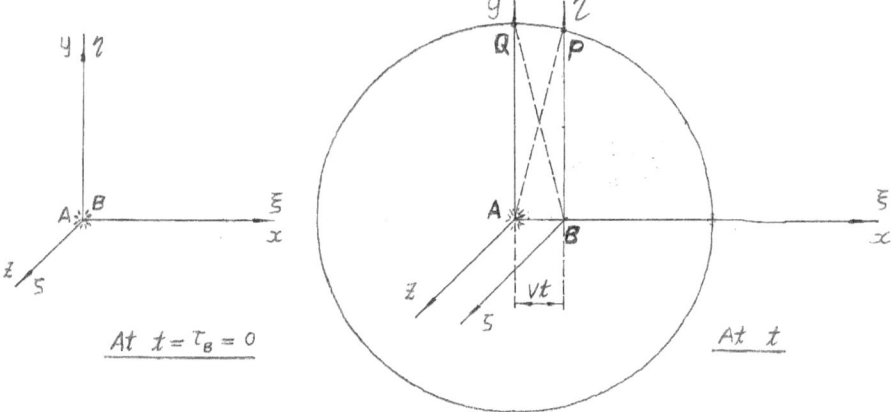

Fig.7 The comparison between P and Q in LT

Obviously the disorder is caused by the two deadlocks of Lorentz space and Lorentz systemic time, as well as the local-time part.

Recall the original case of the two systems under standard configuration where the initial light flash emitted from coincided origins A and B at zero initial instant $t_o = \tau_{BO} = 0$ (Fig.7). The path of the light ray which arrives at point Q and at time t is $[BQ]_B = c\tau_Q = \beta ct$ in System B [from LT, $x_Q = 0$ and $\tau_Q = \beta(t - x_Q v/c^2) = \beta t$], while from the path triangle $[\triangle AQB]_A$ in System A, $[AQ]_A = ct$, $[AB]_A = vt$ and hence $[BQ]_A = t\sqrt{c^2 + v^2}$. We see at once $[BQ]_B \neq [BQ]_A$, yet $[BQ]_B$ and $[BQ]_A$ have the same direction angle as: $arc\ tan([AQ]_A/[AB]_A) = arc\ tan(c/v)$ contradictorily.

Only those, like point P, can avoid such a contradiction. The path triangle in System A for the light ray arriving at P and at t is $[\triangle APB]_A$ which has side $[AP]_A = ct$ and side $[AB]_A = vt$ and hence side $[BP]_A = t\sqrt{c^2 - v^2}$, while in System B, from LT, $x_P = vt$ and $\tau_P = \beta(t - x_P v/c^2) = t/\beta$, and hence $[BP]_B = c\tau_P = ct/\beta = t\sqrt{c^2 - v^2}$. So, we obtain $[BP]_A = [BP]_B$. This is because $[BP]_B$ is perpendicular to the x-ξ axis (lying on the geometric central plane) and would not be dilated by LT and hence could exactly coincide with $[BP]_A$ in the overlapping space. It is this special path triangle $[\triangle APB]_A$ that was always used by SR to make general explanation in the thought experiments unfairly.

Since $[BP]_A/t = \sqrt{c^2 - v^2}$, this is why Einstein had said in [1] that '. . . *light is always propagated along these axes* (namely the η – and the ζ-axis), *when viewed from the stationary system, with the velocity* $\sqrt{c^2 - v^2}$, . . .'. Here we can see how the Postulate causes the two systems having different systemic time. In spite of the fact that the velocity on $[BP]_A$ is $[BP]_A/t = \sqrt{c^2 - v^2}$ in System A, the Postulate requires that the velocity on the coincided length $[BP]_B$ must be $[BP]_B/\tau = c$ in System B (τ should be the systemic time in System B corresponding to t in System A). That is why the systemic time rate between the two systems results in $\tau/t = \sqrt{c^2 - v^2}/c = 1/\beta$ or $\tau = t/\beta$ as a part of Lorentz time as shown in the (15).

By the way, we can see that the points on the surface of the sphere in System B (such as point P and Q) are of different Lorentz time if they have different coordinate ξ, because of their different local-time part. Don't you think it is unreasonable that a wavefront sphere could be formed at a different moment, do you? While in System A, all points are at the same instant t. The situation being different between the two systems reflects the so-called 'relativity of simultaneity' in which the local-time plays a main role. But Einstein claimed it in his thought experiment even with no sight of local-time (see Section 5).

Since LT has unsolvable contradictions in kinematics, it can be anticipated that more defects of LT in dynamics could be found. For simplicity we will neglect the problem with the direction of the velocity in Lorentz System B but focus on its unfitted magnitude only. We should always remember that the velocity vector (and other relevant relativistic vectors) in Lorentz System B contains an unsolvable contradiction between its direction and its magnitude.

Suppose the two systems are inertial and a third party P of mass m has a net force **F** exerted on it and undergoes accelerating rectilinear motion. By applying Newton's second law, we can derive that the acceleration of P in both systems should be equal:

$$\boldsymbol{a_B} = \boldsymbol{F}/m = \boldsymbol{a_A} \tag{18}$$

where $\boldsymbol{a_A} = \dfrac{d}{dt}(\boldsymbol{u})$ that is the acceleration of Body P in inertial System A, and $\boldsymbol{a_B} = \dfrac{d}{d\tau}(\boldsymbol{w})$ that is the acceleration of Body P in inertial System B. The velocity of Body P is $\boldsymbol{u}[u_x,\ u_y,\ u_z]$ in System A and is $\boldsymbol{w}[w_\xi,\ w_\eta,\ w_\zeta]$ in System B.

However, we will prove that, unlike under GT, it actually arrives at $\boldsymbol{a_B} \neq \boldsymbol{a_A}$ under LT.

By differentiating the three equations:

$$w_\xi = (u_x - v)/N \quad w_\eta = u_y/\beta N \quad w_\zeta = u_z/\beta N \quad \text{(where } N = 1 - u_x v/c^2) \tag{8a}$$

and noting $\dfrac{dt}{d\tau} = 1/\beta N$, we can get:

$$a_\xi = \frac{d}{d\tau}w_\xi = a_x/\beta^3 N^3 \neq a_x = \frac{d}{dt}u_x \tag{19a}$$

$$a_\eta = \frac{d}{d\tau}w_\eta = (a_y/\beta^2 N^2) + a_x u_y v/c^2\beta^2 N^3 \neq a_y = \frac{d}{dt}u_y \tag{19b}$$

$$a_\zeta = \frac{d}{d\tau}w_\zeta = (a_z/\beta^2 N^2) + a_x u_z v/c^2\beta^2 N^3 \neq a_z = \frac{d}{dt}u_z \tag{19c}$$

We see at once that indeed $\boldsymbol{a_B}\ [a_\xi,\ a_\eta,\ a_\zeta] \neq \boldsymbol{a_A}\ [a_x,\ a_y,\ a_z]$ (not to mention the problem with their direction angles). The result tells that LT is incompatible with 'the principle about inertial systems'. It can be anticipated that the Law of Conservation of Momentum only holds in the Galilean inertial system but not in the Lorentz system. In other words, Lorentz system is unable to be an inertial system. Yet Einstein had presumed the two systems must be both inertial at the very beginning in SR.

In this section, we have explored Lorentz system by virtue of the Galilean system as a 'stepping stone' and found LT's two 'deadlocks' can never be unlocked. Plus, with other defects on kinematics and dynamics, there is definitely no way for LT, with full of contradictions, to gain 'real meaning'.

4. SR IS DOMINATED BY A SERIOUS CONCEPTUAL ERROR OF 'APPLYING LT WRONGLY ON SECOND PARTY BODY'

In spite of the existence of the two 'deadlocks' in LT, SR insisted the application of the fallacious LT and alleged an *Assertion* that 'moving rod shortens, moving clock slows'.

Common sense immediately judges that such things were absolutely impossible to happen in the absence of dynamic effects.

But common sense was unable to win over SR's 'revolutionary idea'. Thus, we have to ascertain what's wrong with the Assertion in theory.

In 1905 paper [1], Einstein first made the Assertion with two examples in '§4 Physical meaning of the equations obtained in respect to moving rigid bodies and moving clocks'.

Einstein wrote the first example as follows (notice that in all the quotations, some symbols may have been adjusted for more understandable and β is always the Lorentz factor):

'We envisage a rigid sphere of radius R, at rest relatively to the moving system H, and with its centre at the origin of coordinates of H. The equation of the surface of this sphere moving relatively to the system K with velocity v is:

$$\xi^2 + \eta^2 + \zeta^2 = R^2 \qquad (a)$$

'The equation of this surface expressed in x, y, z at the time t = 0 is:

$$\beta^2 x^2 + y^2 + z^2 = R^2 \qquad\qquad (b)$$

'A rigid body, which, measured in a state of rest, has the form of a sphere, therefore has in a state of motion – viewed from the stationary system – the form of an ellipsoid of revolution with the axes R/β, R, R.

'. . . For v = c, all moving objects – viewed from the stationary system – shrivel up into plain figures . . .'

Here, Einstein took $\xi = \beta x$, $\eta = y$, and $\zeta = z$ (at $t = 0$) from LT(8) and, by substituting them into (a), derived (b). But it was a wrongdoing, because (a) was not the equation of a light wavefront sphere ($R = constant \neq c\tau$). There was no reason to substitute LT (8) into (a). Perhaps Einstein recognised this error. He replaced the rigid sphere with a rod in 1916 (Reference [2]).

Einstein continually wrote the second example in §4 of [1]:

'Further, we imagine one of the clocks which are qualified to mark the time t when at rest relatively to the stationary system and the time τ when at rest relatively to the moving system to be located at the origin of the coordinates of k (it was the moving system [τ, ξ, η, ζ] in the (1)), and so adjusted that it marks the time τ. What is the rate of this clock when viewed from the stationary system?

'Between the quantities x, t, and τ, which refer to the position of the clock, we have, evidently, x = vt and

$$\tau = \beta(t - x\, v/c^2)$$

'Therefore,

$$\tau = t/\beta = t - (1 - 1/\beta)t$$

'whence it follows that the time marked by the clock (viewed in the stationary system) is slow by (1 – 1/β) seconds per second, or – neglecting magnitudes of fourth and higher order – by $(vc)^2/2$.

'From this there ensues the following peculiar consequence. If at the point A and B of K (it was the stationary system in the (1)) there are stationary clocks which, viewed in the stationary system, are synchronous and if the clock at A is moved with the velocity v along the line AB to B, then on its arrival at B, the two clocks no longer synchronise, but the clock moved from A to B lags behind the other which has remained at B by $t(v/c)^2/2$ (up to magnitudes of fourth and higher order), t being the time occupied in the journey from A to B.

'. . . If one of two synchronous clocks at A is moved in a closed curve with constant velocity until it returns to A, the journey lasting t seconds, then by the clock which has remained at rest, the travelled clock on its arrival at A will be $t(v/c)^2/2$ second slow . . .'

One thing funny should be pointed out here. The clock that has not been moved is representing the stationary system's origin and recording Galilean time t. The clock that has travelled is representing the moving system's origin and recording Lorentz time τ. Under the standard configuration of LT, the two origins coincide at zero initial instant $t = \tau = 0$. But when they meet once again at the newly coincided point, the stationary clock records Galilean time as t, while the moving clock would gain two conflicting Lorentz times: (t/β) and (βt), because according to LT(8), the Lorentz time τ always is (t/β) at the moving system's origin ($\xi \equiv 0$) and always is (βt) at the point of the moving system coinciding with the stationary system's origin ($x \equiv 0$). This contradiction tells us that LT cannot be trusted. So it was unexpected by Einstein that the example he wanted to explain his theory turned out to negate the theory.

We have seen that Einstein had applied the (14) and the (16) to claim the Assertion in the two examples, but the logical way to make the Assertion is contrary between the two examples.

In the second example, he let a stationary system clock record Galilean time Δt and let a moving clock record Lorentz time $\Delta\tau$. Then he directly claimed 'moving clock slows' as $\Delta\tau < \Delta t$ according to $\Delta\tau = \Delta t/\beta$ (16).

But in the first example, he fixed the body's size $\Delta\xi$ in the moving system as 'the resting length' (along the moving direction) and let the body's size $\Delta x = \Delta\xi/\beta$ (14) in the stationary system be treated as 'the moving length' of the body. Then he claimed 'moving body shortens' as $\Delta x < \Delta \xi$. Actually $\Delta \xi$ should be the moving length viewed from the stationary system and Δx should be the resting length measured from the stationary system. Einstein had made things upside down, such that dilating Lorentz space (under projection) had been wrongly treated as contracting. Perhaps he intended to make his Assertion to be in line with the famous 'Lorentz Contraction' in M-M Experiment (see Section 6). He probably recognised that it would be a problem with SR if the two parts of the Assertion was not to be claimed in a unified way. So he tried, but failed, to claim 'moving clock slows' in the same way as claiming 'moving body shortens', which can be seen in section xii of [2].

Claiming 'moving clock slows' with $\Delta \tau = \Delta t/\beta$ while claiming 'moving body shortens' with $\Delta x = \Delta \xi/\beta$ would lead to an unreasonable result that the velocity of the moving system relative to the stationary system would be wrongly expressed as

$$v = \Delta x/\Delta \tau = \Delta \xi /\Delta t$$

This mistake often can be seen in SR's calculation for space travel. It obviously violates the concept of the relative velocity between two systems.

This mistake was wrong not only on crossing the coordinates between the two systems but also on mixing third party with second party in LT. All of the Δx and $\Delta \xi$, as well as the $\Delta \tau$ and Δt, appearing in the above equation originally come from the equations of LT(8) and hence represent the displacement or the travelling time of third party body. But, in the above equation, they were used to calculate, not the velocity of third party (u or w), but the relative velocity (v) between the two systems which was always the velocity of second party relative to first party. That was a serious conceptual error of 'confusing second party body with third party body' or, in other words, 'applying LT wrongly on second party'.

A more obvious evidence can be seen in '§10. Dynamics of the (Slowly Accelerated) Electron' of the [1]. On the page 158 Einstein wrote that '. . . Since the electron is supposed to accelerate slowly, and consequently cannot emit any energy in the form of radiation, the energy taken from the electrostatic field must be equated to the kinetic energy W of the electron. Bearing in mind that the first of equation (A) [i.e. $\frac{d^2x}{dt^2} = \frac{\varepsilon}{\mu} \frac{1}{\beta^3} X$] holds throughout the entire process of motion, we obtain

$$W = \int \varepsilon X dx = \int_0^v \beta^3 v dv = \ldots \ldots '$$

You can see that Einstein had mixed the x-component of third party velocity $u_x = \dfrac{dx}{dt}$ with second party velocity $v = \dfrac{dx}{dt}$. The above equation should be rectified as 'W $= \int_0^{u_x} \beta^3 u_x du_x = \ldots \ldots$'. The result would be completely different.

Such a conceptual error of 'applying LT wrongly on second party' dominates the whole of SR, especially the Assertion of 'moving rod shortens, moving clock slows'. Following is the explanation of why the Assertion is hiding the error.

There are two different cases we should distinguish in dealing with motion issues.

In *the case of 'single motion'* where there is no third party body but only second party body moving with *v* relative to the stationary system as the reference, there is no need to employ any transformation formulae and no need to involve in the moving system [τ, ξ, η, ζ] which only opens for third party exclusively. The second party body just simply experience the space and time of the stationary system [t, x, y, z], no matter where it is sitting on system [τ, ξ, η, ζ] (as learnt in Section 1, the position between a body and its attaching system does not matter, as long as they are maintained at rest relatively to each other). The first party body carrying the stationary system as the reference for motion is useless in practice and may be hidden.

In *the case of kinematic transformation,* where, in addition to the case of 'single motion' that a second party body is moving with respect to the stationary system, there are third party bodies in motion relative to both the stationary system [t, x, y, z] carried by the first party body, and the moving system [τ, ξ, η, ζ] carried by the second party body. On one hand, the second party body opens up system [τ, ξ, η, ζ] as a reference for third party's motion; on the other hand, it itself experiences the space and time of the stationary system [t, x, y, z], as the same situation as in the case of "single motion", no matter where it is sitting on system [τ, ξ, η, ζ]. Only third party bodies would experience the space and time of both system [t, x, y, z] and system [τ, ξ, η, ζ] simultaneously. Thus, transformation formulae are needed for the third party bodies.

In the two examples in §4 of [1], the rigid sphere or the clock was *resting* in the moving system; they surely belong to the second party. They only experience normal Galilean space and time of the stationary system and hence would not shorten or slow, respectively. The two examples are cases of 'single motion' of no third party and no need for applying LT. Therefore, the Assertion is definitely not true.

Einstein wrongly thought that the rigid sphere or the clock sitting on the moving system must experience the space and time of the moving system. He failed to recognise the nature of second party and distinguish it from third party in transformation of motion. This is not a defect of LT, but a serious conceptual error of 'applying LT wrongly on second party body' in the Assertion.

Note that even for a third party body, the Assertion cannot be claimed. Here are the reasons:

- Third party is always the moving bodies, not the resting bodies, relative to the two systems. There is no comparison between moving state and resting state for them.

- Since Lorentz time τ varies between different ξ (caused by local-time part), third party body with size moving in Lorentz system would have many different velocities simultaneously. Actually, such phenomenon could not happen. Thus third party body should be handled as point-mass or particle without size, and distortion issues should be dismissed.

- A third party clock would experience Galilean time t and Lorentz time τ simultaneously. It is a mission impossible for the clock to record two different times.

We see at once that the Lorentz system is actually immeasurable in practice. Under no circumstance can a moving particle of third party simultaneously mark two different lengths of an interval at a fixed Galilean time (14) or a moving clock of third party simultaneously record two different times at a fixed spatial spot of Lorentz system (16). Therefore, Einstein's passion to measure Lorentz space and Lorentz systemic time by means of rod and clock is absolutely unrealistic. Only calculation of LT associated with the missing light source system could determine Lorentz system, including the missing Lorentz local-time that will be discussed in the next section.

There have been exciting paradoxes explained by SR alongside the wrong Assertion. They all are a case of 'single motion' but have been described with the conceptual error of 'applying LT wrongly on second party body' (not to mention other problems similar to those in the two examples in §4 of [1]).

In the paradox of space travel, the spaceship is the second party body travelling in Earth's external space as the stationary Galilean system. The situation of no third party is so simple that there is no need to bother with any transformation, neither LT nor GT. How can the travelling time shorten? Yet to dream a shorter distance, too. What wishful thinking!

In the famous Twin Paradox, things are just like the clock moving along a circle in the second example in §4 of [1]. Upon the returning twin's arrival on Earth at time t, what time would his watch record: (t/β) or (βt)? This contradiction cannot be solved with SR. But don't be bothered by this, for the situation of 'twin paradox' is similar to the space travel: no third party, no LT. Hence, the watch would also show time t meaning no age difference between the twins (Even if Einstein's General Theory of Relativity could determine which twin would age less in the Twin Paradox, it cannot be seen as a success. Rather, it actually confirms SR's failure).

The Assertion also claims that the prolonged lifespan of the muons detected in cosmic ray showers is due to time dilation. Actually, the muons were just the second party in the case of 'single motion' and would be in no way able to experience the so-called time dilation. The reason why muons have a much longer lifespan in high speed motion than when at rest may be searched from the dynamics viewpoint that the higher the kinetic energy, the longer the decay. SR had mixed the concept of lifespan with time flux.

So far, our criticism on the Assertion just focuses on the conceptual error of SR, not the defects of LT. If further considered the falseness of LT, the Assertion would lose ground thoroughly.

The conceptual error of 'applying LT wrongly on second party body' also dominates the so-called relativistic mechanics.

In the '§10. Dynamics of the (Slowly Accelerated) Electron' of [1], Einstein let the moving System B $[\tau, \xi, \eta, \zeta]$ sticking with the electron together in the motion relative to the stationary System A $[t, x, y, z]$ with velocity v at a particular moment. At the moment, the electron is at rest ($w = 0$) and has the rest mass m_0 relative to System B, but it is also accelerating with acceleration $\mathbf{a_B}$ $[a_\xi, a_\eta, a_\zeta]$ under the influence of force $\mathbf{F}$ $[F_{x-\xi}, F_{y-\eta}, F_{z-\zeta}]$. Then he applied LT on the electron. As a result, he introduced new concepts of:

$$\text{Longitudinal mass} = m_0 / [\sqrt{1-(v/c)^2}]^3 = m_0 \beta^3$$

$$\text{Transverse mass} = m_0 / [1-(v/c)^2] = m_0 \beta^2$$

for the electron (relative to System A) in order to preserve the relation of 'Mass x Acceleration = Force'.

Here, we give a general approach different to that in [1]. In the last section, we have shown LT's incompatibility with Newton's laws by the (19 a, b, c). As LT is reversible in form between the two systems, we can easily rewrite the (19 a, b, c) as:

$$a_x = a_\xi /\beta^3 N_B^3$$

$$a_y = (a_\eta /\beta^2 N_B^2) - a_\xi w_\eta v/c^2\beta^2 N_B^3$$

$$a_z = (a_\zeta /\beta^2 N_B^2) - a_\xi w_\zeta v/c^2\beta^2 N_B^3$$

where $N_B = 1 + w_\xi v/c^2$.

Now, since $w_\xi = w_\eta = w_\zeta = 0$ at the particular moment, we have $N_B = 1$ and

$$a_x = a_\xi /\beta^3 \qquad a_y = a_\eta /\beta^2 \qquad a_z = a_\zeta /\beta^2$$

Then, noting that $(F_{x-\xi} /a_\xi) = (F_{y-\eta} /a_\eta) = (F_{z-\zeta} /a_\zeta) = m_0$ in System B, we derive the longitudinal mass m_x and the transverse mass m_y or m_z for the electron in System A:

$$m_x = F_{x-\xi} /a_x = \beta^3 F_{x-\xi} /a_\xi = m_0 \beta^3$$

$$m_y = F_{y-\eta} /a_y = \beta^2 F_{y-\eta} /a_\eta = m_0 \beta^2$$

$$m_z = F_{z\text{-}\zeta}/a_z = \beta^2 F_{z\text{-}\zeta}/a_\zeta = m_0\beta^2$$

One can see that these two new concepts of mass were introduced under a very special condition at a particular moment and hence would not have generality; if constantly doing it so, System B would become an accelerating system ineligible to take part in LT.

Actually, the situation was a case of 'single motion' of a 'slowly accelerated electron' in reference System A, involving no third party's transformation. The electron carrying System B was just a second party to System A. Einstein had wrongly applied LT on this electron and given the two new masses to it. Consequently, applying these new masses for the electron in System A would result in a_B $[a_\xi\ a_\eta\ a_\zeta] = a_A$ $[a_{x'}\ a_y,\ a_z]$ indeed, but how about the force in System A? What are such things like $(a_\xi\ m_0\beta^3)$, $(a_\eta\ m_0\beta^2)$, and $(a_\zeta\ m_0\beta^2)$? Can we name them 'relativistic force' in System A?

It seemed Einstein had really had such a revolutionary idea as he wrote in [1] (page 157) that 'Of course, with a different definition of force and acceleration, we would obtain different values for these masses'. If so, things would have gone farther beyond the reality.

Another attempt of covering up LT's incompatibility with Newton's laws was first to make a new law of conservation of 'relativistic momentum'. It started from the general definition of *relativistic mass*

$$m = m_0/\sqrt{1-(V/c)^2} \qquad\qquad (20)$$

where V is the full velocity of the object relative to a reference system and m_0 is the rest mass of the object measured in another system where the body is resting.

Formula (20) shows that the relativistic mass m varies as relative velocity V changes: $m = f(V)$. In other words, m is relative to the reference system. When $V = 0$ (the body is at rest in the reference system), $m = m_0$ (the smallest), and hence m_0 is named 'the rest mass' of the object. The rest mass m_0 is an inherent character of the body and is equalled to its original mass in classical mechanics.

Then the *relativistic momentum* in direction *i* can be expressed as

$$p_i = m V_i = m_o V_i / \sqrt{1 - (V/c)^2} \qquad\qquad (21)$$

It is important to note that V comes from the definition of the relativistic mass and must be the full velocity of the body while V_i is just its component in the particular direction i.

It had been proved that, with the aid of the concept of relativistic mass, LT would be compatible with a new law of conservation of relativistic momentum superseding the classical mechanics law of conservation of momentum (see Reference [6]).

But the definition of momentum vector $\boldsymbol{p} = m\boldsymbol{V}$ is originally stipulated that $\boldsymbol{p}$ has the same direction as $\boldsymbol{V}$ and its value p is directly proportional to v. Hence, their constant ratio m must be independent of v and is an inherent character of the object, as the mass, representing the object's inertia. Thus, the definition of relativistic mass as a function of V is unreasonable. It was just like playing a definition game.

Max Born had a good try in shifting the artificial definition of relativistic mass into a natural relation (see [4], page 268–271). He investigated the completely inelastic collision of two equal spheres in a reference system. He let the two spheres representing two systems without any third party body; then applied LT on the spheres and derived the (20) for the relativistic mass.

But actually, the spheres were both second party to each other and there should have been no kinematic transformation involving it. Thus, it had clearly exposed that, like the Assertion, the concept of relativistic mass was a product with conceptual error of 'applying LT wrongly on second party body' indeed. Therefore, the definition of relativistic mass and the new law of conservation of relativistic momentum are unrealistic.

Notice that the concept of relativistic mass could lead to a simple statement that a 'moving mass increases', which, together with the Assertion of 'moving rod shortens, moving clock slows', were three famous arguments in SR. They have been expressed as three formulae in high school syllabus of physics:

$$L = L_0 \sqrt{1 - (v/c)^2} \qquad t = t_0 / \sqrt{1 - (v/c)^2} \qquad m = m_o / \sqrt{1 - (V/c)^2}$$

where L_o, t_o, and m_o are respectively the object's rest length, time, and mass measured in the System B, where the object is resting, while L, t, and m are respectively the object's moving length, time, and mass measured in the System A with velocity v or V relative to the object.

Actually, it is the object carrying System B together and doing a simple motion with v or V relative to System A. Since the V in the third formula actually has the same meaning as the v in the first and the second formula, thus these three formulae are all containing the Lorentz factor $\beta = 1/\sqrt{1-(v/c)^2}$ and clearly indicate that they must all come out of the conceptual error of 'applying LT wrongly on second party body'.

Can the concept of relativistic mass help SR to cover up LT's incompatibility with Newton's laws? No.

Newton's second law had a general form

$$\boldsymbol{F} = \frac{d}{dt}(m\boldsymbol{V})$$

For the case where the net force is in line with rectilinear motion:

$$F = \frac{d}{dt}(mV) = V\frac{d}{dt}(m) + m\frac{d}{dt}(V) \tag{22}$$

Let the rectilinear motion be along a coordinate axis in reference system with $V_i = V$, using the concept of relativistic mass (21), it would yield that:

$$F = \frac{d}{dt}[m_o V/\sqrt{1-(V/c)^2}\,] = m'\frac{dV}{dt}$$

where $m' = m_o/[\sqrt{1-(V/c)^2}\,]^3$. It seemed like good news for SR that the m' was the same as the longitudinal mass for the slowly accelerated electron. But this mass was deduced only when the rectilinear motion was along a coordinate axis in a reference system. Otherwise, the rectilinear motion would have three components in general:

$$F_x = \frac{d}{dt}[m_o V_x/\sqrt{1-(V/c)^2}\,] = m'\{[1/(1-v^2/c^2)]\frac{d}{dt}(v_x) + (v_x v/c^2)\frac{dV}{dt}\}$$

$$F_y = \frac{d}{dt}[m_o V_y/\sqrt{1-(V/c)^2}\,] = m'\{[1/(1-v^2/c^2)]\frac{d}{dt}(v_y) + (v_y v/c^2)\frac{dV}{dt}\}$$

$$F_z = \frac{d}{dt}[m_o V_z/\sqrt{1-(V/c)^2}\,] = m'\{[1/(1-v^2/c^2)]\frac{d}{dt}(v_z) + (v_z v/c^2)\frac{dV}{dt}\}$$

What a horrible mess that more possible relativistic masses could be introduced! Moreover, different coordinate systems have yielded conflicting results. These problems are caused by the full velocity V appearing in the component equations and indicate that the concept of relativistic mass cannot be trusted. It can be anticipated that the problems would be more serious if there was further transformation from this reference system to another moving system under LT.

We need not to step into the mess further, as other defects of LT would certainly cause the entire SR, including the relativistic mechanics, to collapse.

It should be noted that the equation of $E = mc^2$ is unnecessarily linked with relativistic mechanics, because it can be derived in other ways (see page 283–286 of [4], a proof due to Einstein himself without making use of SR). So SR's problems have no effect on it.

In this section, we have seen that Einstein carried on explanation and application of LT in a very strange and improper way, so that many paradoxes had been generated. But all were actually a conceptual error of 'applying LT wrongly on second party'. The uncertain world with relative space and time created by SR has no theoretical support and hence must be unrealistic.

5. WE ARE ALL DECEIVED BY THE 'TIME' OF NO FLUX

In last section, we have focused on the so-called space dilation from $\Delta \xi = \beta \Delta x$ (14) and systemic time shortening from $\Delta \tau = \Delta t / \beta$ (16) in the Lorentz system and found that there is no way to resolve these two deadlocks. In this section, we will focus on the missing Lorentz local-time $(-\frac{V}{C^z}\xi)$. As a part of Lorentz time, local-time $(-\frac{V}{C^z}\xi)$ is a mystery that SR never gave a thought to.

We shall first explore more on how the landscape of Lorentz local-time in System B is.

In Section 3, we have learnt that the Lorentz time τ (15) consists of systemic time (t/β) and local-time $(-\frac{V}{C^z}\xi)$. At the initial instant $t = 0$ in System A, the systemic time part of Lorentz time would be zero for the entire System B, while local-time has been already preset on every point, differing between η-ζ planes. The preset local-time varies linearly as ξ varies. We may call this phenomenon 'the landscape of local-time'.

Once preset, local-time at every fixed point (certain ξ) in Lorentz System B never changes. The value of the preset local-time can be determined only by calculation. It is naive to think of measuring local-time with a clock. This is another reason why Lorentz time is immeasurable.

Under the standard configuration for LT, the zero local-time plane coincides with the geometric central plane at $\xi = 0$, where Lorentz time always has just the systemic part as $\tau = t/\beta$, which presents the present time in Lorentz System B corresponding to the current moment t in Galilean System A.

The Lorentz space is divided by the zero local-time plane into two zones: the zone with plus local-time ($-\dfrac{V}{C^z}\xi > 0$) is always at the time before the present time (t/β), while the zone with minus local-time ($-\dfrac{V}{C^z}\xi < 0$) is always at the time after the present time (t/β). The further the place is apart from the zero local-time plane, the further its time is apart from the present time.

So, a third party (including the flash) moving in Lorentz System B would sometimes travel in the past, sometimes travel in the future, due to the landscape of local-time. For the wavefront sphere of the flash, only the intersection of its surface with the zero local-time plane is just arriving now at the present, while one part of its surface on one side of the zero local-time plane has already arrived on an earlier time of the past, another part of its surface on another side will arrive on a later time of the future.

What an unthinkable and unacceptable situation made by the local-time!

Another point should be made. Since the Lorentz local-time part differs from ξ to ξ, a solid body would have different Lorentz time on different parts of the body and hence cannot have a united translational velocity for the whole body in Lorentz space. Therefore, in Lorentz space, a solid body has to be handled as a point-mass with no size, and hence it should not involve in any shape distortion that had been asserted by SR.

Relativity of simultaneity was the first selling point of SR. It means that 'two events that are simultaneous when observed from some particular coordinate system can no longer be considered simultaneous when observed from a system that is moving relative to that system' (quoted from Einstein's 1905 SR paper [1]). Being the theoretical Bible of SR, LT should have been used to explain this important issue. Instead, SR just uses thought experiment (the 'Train Paradox') to make people believe that relativity of simultaneity is a natural phenomenon under the Postulate of 'absolute velocity of light'.

Actually, local-time is the sole player in relativity of simultaneity. From LT (8), we can obtain time difference between two spatial spots as

$$\Delta\tau = -\frac{V}{C^z}\Delta\xi \quad \text{(condition: } t = constant\text{)} \tag{23}$$

$$\Delta t = \frac{V}{C^z}\Delta x = \beta\frac{V}{C^z}\Delta\xi \quad \text{(condition: } \tau = constant\text{)} \tag{24}$$

Formula (23) tells that when two events happen at the same Galilean time t in System A, they appear in System B at different Lorentz time, which is the difference of local-time between them.

Formula (24) tells us that when two events happen at equal Lorentz time τ = t/β − $\frac{V}{C^z}$ ξ in System B, they appear in System A at different Galilean time due to the difference, with a factor of β, of local-time between them.

Let's have a look at the Lorentz time difference between two typical points P and Q on the surface of the flash wavefront sphere in Fig.(7) (see Section 3).

From LT, for point P: $x_P = vt$ and $\tau_P = \beta(t - x_P v/c^2) = t/\beta$ while for point Q: $x_Q = 0$ and $\tau_Q = \beta(t - x_Q v/c^2) = \beta t$.

We obtain $\tau_Q - \tau_P = \beta t - t/\beta = \beta t v^2/c^2 = -\frac{V}{C^z}\xi_Q$ since $\xi_Q = \beta(x_Q - vt) = -\beta vt$, as $x_Q = 0$. This result demonstrates that:

1. Even though it is at an identical instant t in System A, there is a 'time difference' in System B between two points of different ξ, such as P and Q, reflecting the so-called relativity of simultaneity.
2. The time difference in System B between two points always equals their local-time difference, which is unchangeable once preset.
3. It is such local-time difference (together with space dilation and systemic time shortening) to help light rays in all radical directions from origin B travelling with c in System B, and that is how LT could be able to support the Postulate of 'absolute velocity of light'.

It cannot be clearer that it is the local-time part of Lorentz time that produces SR's relativity of simultaneity, which the systemic time part (t/β) of Lorentz time has nothing to do with. The thought experiment of Einstein's Train Paradox had completely missed the point and had given a misleading explanation for the so-called relativity of simultaneity. LT should have been treated as the Bible for SR to make authentic explanation of everything. But Einstein put it aside and carried on the business of SR in a unique and strange way.

It is also obvious that local-time is a dispensable part of Lorentz time, although it has a size very little (particularly when $v \ll c$). It plays a vital role in supporting the Postulate. The importance of local-time can be seen in the sight that it appears on the right-hand side of the equation (9). Without local-time part of Lorentz time, equation (9) could not hold.

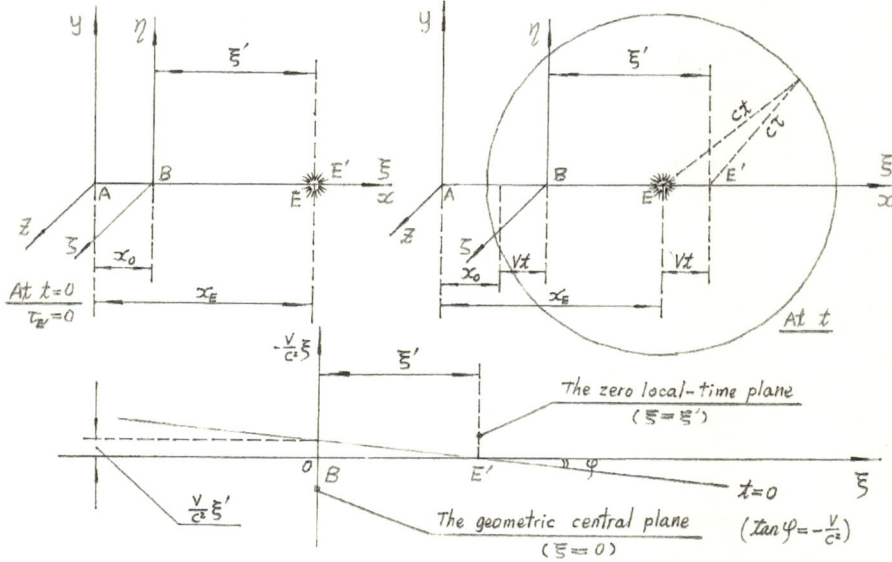

Fig.8 LT (Under non-standard configuration)

Now we turn to discuss the phenomenon that different configurations would make different landscape of local-time and hence different forms of LT.

Consider a situation under a non-standard configuration as shown in Fig.8, where the light source is seated at the point E in System A with a distance of x_E from origin A and the point E has coordinate $\xi = \xi'$ in System B at zero initial instant $t = 0$ when the origin B already has a displacement of x_0 relative to origin A. The equation of the wavefront sphere at time t for System A and for System B will respectively be:

$$(x - x_E)^2 + y^2 + z^2 = c^2t^2 \text{ and } (\xi - \xi')^2 + \eta^2 + \zeta^2 = c^2\tau^2$$

Under such non-standard configuration, LT will take a different form:

$$\xi - \xi' = \beta[(x - x_E) - vt] \quad \text{i.e. } \xi = \beta(x - vt) + \xi' - \beta x_E$$

$$\tau = \beta[t - \frac{V}{C^2}(x - x_E)] = t/\beta - \frac{V}{C^2}(\xi - \xi')$$

As a result, the central plane is still at $\xi = 0$, because x_E and ξ' are constant so that $\Delta\xi = \beta\,\Delta x$ retained. But the zero local-time plane is located newly at $\xi = \xi'$, because local-time has been altered as $[-\dfrac{V}{C^z}(\xi - \xi')]$. Now when at the initial instant $t = 0$, the origin B at $\xi = 0$ already has Lorentz time $\tau = \dfrac{V}{C^z}\xi'$, while every point in system B has Lorentz time containing local-time with an extra part of ($\dfrac{V}{C^z}\xi'$) in comparison to the situation where the zero local-time plane coincides with the central plane under standard configuration (notice that local-time difference between two locals, $\dfrac{V}{C^z}\Delta\xi$, is unchangeable).

It is not hard to realise that the central plane is always at origin B ($\xi = 0$), while the zero local-time plane is always at the point of ξ', that is the point in System B coinciding with System A's light source at the initial instant when the first flash is emitted.

Therefore, if $\xi' \neq 0$, the zero local-time plane would be apart from the central plane. Consequently, a different landscape of Lorentz local-time would appear and hence a different Lorentz system would emerge. That means LT relies upon its configuration due to the existence of local-time. Such phenomenon is unacceptable in theory.

We have seen that LT not only has dependency on light source system but also has dependency on configuration. This fact of uncertainty reflects that LT is just a particular case relying on a particular light flash as a third party under the Postulate of 'absolute velocity of light' and hence should not be taken as a general law of transformation.

However, carefully contrasting with Lorentz systemic time could shockingly expose that local- time has actually no property of flux.

Where can we see the fact that the local-time is a false part of Lorentz time?

1. As we have seen in the preceding section, the increment of the whole Lorentz time $\Delta\tau$ at an arbitrary point fixed in System B, corresponding to the lapse of universe time Δt, is

$$\Delta\tau = \Delta t\,/\,\beta \qquad (\text{condition: } \xi = constant)$$

There is only the increment of the systemic time part (t/β) occurring in $\Delta\tau$, but no sight of the increment of local time ($-\frac{V}{C^z}\xi$). That implies that time running has nothing to do with local-time for any fixed point of Lorentz space.

2. Between any two events, the increment of Lorentz time in Lorentz System B, corresponding to the increment of Galilean time t in Galilean System A, is

$$\Delta\tau = \Delta t/\beta - \frac{V}{C^z}\Delta\xi$$

Only the Lorentz systemic time increment is truly the part that fully corresponds to t, while the part of local-time difference ($-\frac{V}{C^z}\Delta\xi$) has no correspondence at all. The Lorentz local-time difference in system B is quite like 'city time difference' between two cities of different longitudes on the globe, which has nothing to do with flight time running during travel on airline but just needs to adjust your watch by hand at arrival of your destination. Thus, Lorentz local-time only has numerical apparent value but no real physical meaning.

3. Local-time is preset at the initial instant and then never lapses afterwards during the universe time t fluxes. Once preset, it would have an unchangeable value at every fixed point forever in Lorentz space in spite of Galilean time (t) and Lorentz systemic time (t/β) fluxing. Unless by calculation, there is no way to measure local-time with a clock directly.

4. Local time comes from the anti-GT assumption that artificially arranges a link between time and space for the deduction of LT. The assumption has no evidential ground to endow space with the property of time flux. Of course, local-time cannot be real.

5. In Fig.7, the two light rays, emitted from the two coinciding origins at zero initial instant $t_0 = \tau_{B0} = 0$ arrive at point P and point Q when the universe time in System A is t while the Lorentz systemic time in System B is (t/β). There is no Lorentz local- time involved at all during the time lapsing.

6. It is local-time that would make the Lorentz time varying on the surface of the wavefront sphere in System B (that apparently helps LT to achieve that all rays of the flash could have a velocity $w = c$, as required by the Postulate). But such unreasonable phenomenon is definitely impossible (the whole wavefront must appear simultaneously). That means local-time would not exist.

This is absolutely a startling discovering that the Lorentz time τ even contains a part of ($-\dfrac{V}{C^z}\xi$) with no sense of time! We all have been deceived by that false local-time.

The consequence is quite serious. No local time, no relativity of simultaneity, and no theoretical support for the Postulate of 'absolute velocity of light' from a fallacious LT. Then the whole theory of SR must collapse at once.

Now, one can believe that LT is purely a mathematical model under 'wishful thinking'. Nature does not listen to big talks. When modern physicists were excited about the revolutionary concepts of SR, about the relativity of simultaneity, about 'moving body contracts, moving clock slows', about the Twins Paradox, about relativistic mechanics, about 'moving mass increases' and about Monkowski's space-time continuum, they had not realised that local-time was deceiving them and they were actually living in a fantastic fanciful world with no objective reality.

Since SR is theoretically untenable, it can be expected that SR would not be attested to experiments. This will be the contents in our next two sections.

6. THE M-M EXPERIMENT IS NOT IN FAVOUR OF SR AT ALL

The 'M-M Experiment' is the Michelson-Morley Experiment for short. Since the M-M Experiment was recognised as a most convincing experiment with high accuracy, SR was very keen to claim support from the result of it.

The purpose of the M-M Experiment was to detect the existence of ether through investigating the influence of the earth's spinning motion on the light rays which were emitted from an apparatus on the ground of Earth.

In the Experiment (Fig.9), the ground of Earth, together with the whole apparatus, was moving with the earth's spinning velocity v relative to 'ether' along the east-west direction. We only needed to focus on the two light rays which travel in a round trip between the central Mirror O and Mirror P along east-west path O-P-O and between the central Mirror O and Mirror Q along south-north path O-Q-O. By default, $L_1 = L_2$, where L_1 and L_2 is the distance OP and OQ respectively.

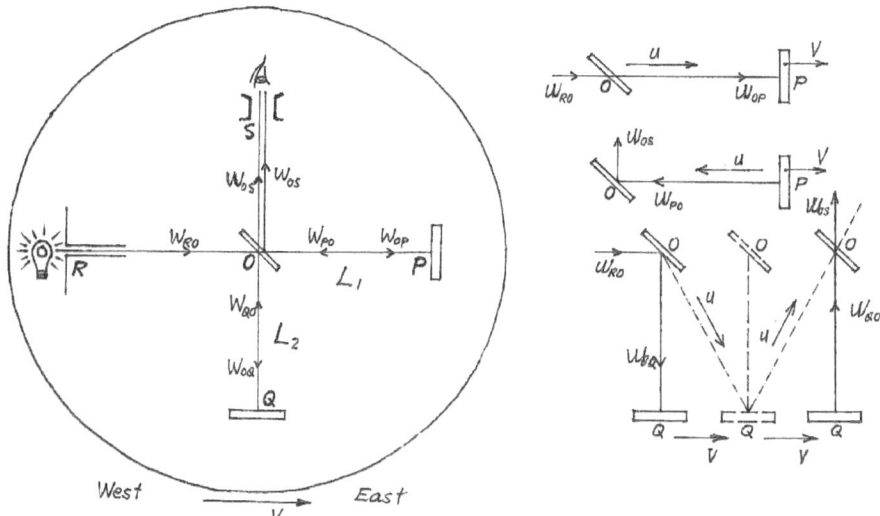

Fig.9 The M-M Experiment (ether was taken as the reference)

According to the Ether Theory, light is always propagated with c relative to the ether. An original analysis using GT together with the Ether Theory to predict the result of the Experiment was as follows.

The ether was taken as a stationary system (reference body), while the apparatus (together with the Earth) was treated as a moving system.

Denote $\boldsymbol{u}$ as the velocity of light relative to the ether, $\boldsymbol{w}$ as the velocity of light relative to the apparatus and $\boldsymbol{v}$ as the velocity of the moving apparatus relative to the stationary ether. According to GT, $\boldsymbol{w} = \boldsymbol{u} - \boldsymbol{v}$ while from the Ether Theory, $u \equiv c$.

On 'trip O-P', $w_{OP} = u - v = c - v$. On 'return trip P-O', $w_{PO} = u + v = c + v$. So the travel time on 'round trip O-P-O' in the moving system (the apparatus) would be

$$(\Delta\tau)_1 = OP/w_{OP} + PO/w_{PO} = \frac{L_1}{c-v} + \frac{L_1}{c+v} = (2L_1/c)\beta^2 \text{ (where } \beta = \frac{c}{\sqrt{c^2-v^2}} \text{)} \quad (25)$$

Meanwhile, on 'trip O-Q', $\boldsymbol{w}$ must lie on OQ so that a right-angled triangle should be formed with the hypotenuse representing $\boldsymbol{u}$. Hence $w_{OQ} = \sqrt{u^2 - v^2} = \sqrt{c^2 - v^2}$. The same situation was on 'return trip Q-O': $w_{QO} = \sqrt{c^2 - v^2}$. So, the travel time on 'round trip O-Q-O' in the moving system (the apparatus) would be:

$$(\Delta\tau)_2 = OQ/w_{OQ} + QO/w_{QO} = \frac{L_2}{\sqrt{c^2-v^2}} + \frac{L_2}{\sqrt{c^2-v^2}} = (2L_2/c)\beta$$

$$\text{(where } \beta = \frac{c}{\sqrt{c^2-v^2}} \text{)}$$

Hence:

$$(\Delta\tau)_1 / (\Delta\tau)_2 = \beta L_1 / L_2$$

Given $L_1 = L_2$ as defaulted, obviously $(\Delta\tau)_1 \neq (\Delta\tau)_2$ and such a difference (as a factor of β) between the two round trips in travel time had been the prediction for the Experiment. Because the time difference would confirm that there indeed is a relative motion between the ether and the earth, and hence would verify the existence of ether.

However, a null result of the Experiment actually turned out being '$(\Delta\tau)_1 = (\Delta\tau)_2$'; even if the apparatus was turned 90°, the result was still the same.

The negative result: '$(\Delta\tau)_1 = (\Delta\tau)_2$', leads to '$\beta = 1$' and hence '$v = 0$', i.e. the relative velocity v between the Earth and the ether is zero. That means either the Earth is at rest in the ether or the ether is absent. But the Earth resting in the ether is unacceptable, because it is impossible for ether to move (spin and orbit) with the Earth. Therefore, the null result of the Experiment had shown that no ether exists at all.

The Ether Theory now faced a difficulty that the concept of ether should be dumped. 'Lorentz and FitzGerald rescued the theory from this difficult by assuming that the motion of the body relative to the ether produces a contraction of the body in the direction of motion, the amount of contraction being just sufficient to compensate for the difference in time . . .' (quoted from [2]). That means L_1 in the direction of v should have contract to equal (L_2/β) in order to conform to the null result '$(\Delta\tau)_1 = (\Delta\tau)_2$'. This suggestion was called 'Lorentz Contraction' (which later became a raw idea for SR).

We see at once that such 'Lorentz Contraction' occurring in the apparatus system is absolutely impossible because the light source in the Experiment was located within the apparatus, and hence, the apparatus (as well as the ground of Earth) was actually carrying the light source Galilean system where no space distortion could happen.

Moreover, a contradiction had been hidden in the original calculation for prediction of the Experiment. Look at the apparatus (Fig.8). As the situation on RO is the same as on OP, $w_{OQ} = w_{RO} = w_{OP} = (u - v) = (c - v)$. Nevertheless, in the original calculation, the velocity on direction OQ had been determined as $w_{OQ} = \sqrt{c^2 - v^2}$. This contradiction between the two results for w_{OQ} indicates that the original calculation for prediction of the Experiment, based on the Ether Theory, is wrong. Hence the suggestion of 'Lorentz Contraction' from the wrong original calculation makes no sense in saving the concept of ether.

It is the Ether Theory requiring $u \equiv c$ that generates the contradiction. Thus, the Ether Theory should be dumped alongside the concept of ether.

After the dissolution of ether, how should we explain the result: '$(\Delta\tau)_1 = (\Delta\tau)_2$'?

Since '$\beta = 1$' and '$v = 0$' in the absence of ether, we get $w_{OP} = w_{PO} = w_{OQ} = w_{QO} = w_{RO} = c$, or $w \equiv c$, and hence $(\Delta\tau)_1 = (\Delta\tau)_2 = 2L/c$. That means the velocity of the light rays, relative to the whole apparatus as well as the ground of Earth, always equals the emitting velocity c, no matter what state of motion the ground is in, as long as the ground could be regarded as an inertial system. Classical mechanics tells us that the motion of anything relative to an inertial system is not affected by the system's own motion with respect to other systems. Thus, Earth's motion (spinning and even orbiting relatively to other planets) has no influence on the travels of light in the apparatus. That is why light's travelling in the two mutually perpendicular round trips would be exactly the same. This is just a simple case of 'single motion', where there is only the apparatus's light source system left in the Experiment with no need to employ LT or GT. Thus the null result is fully understandable with the concept of light source Galilean system in the absence of ether.

Today's students, who have no any historical idea about ether, may think the Experiment is superfluous. However, the null result of the Experiment had made a main contribution to the dissolution of the false concept of ether in history. In addition, it had verified the concept of light source Galilean system and hence partly confirmed the validity of the light's *Emission Theory* of classical mechanics.

The light's Emission Theory is that light moves in the light source system with emitting velocity c, while *simultaneously propagating in the overlapping non-light source systems with velocity different to c*. The first part is the same as the concept of light source Galilean system and should be not contradicted by the Postulate but had been neglected by SR, as we have previously pointed out in Section 2. Only the second part is in conflict with the Postulate.

As there was actually only a single light source system retained in the M-M Experiment, the result of the Experiment was unable to help us making judgement on which of the Postulate and the second part of the Emission Theory was correct.

However, the claim that the Experiment was in favour of LT and SR is definitely not true, as the Experiment had nothing to do with LT at all.

Einstein had written that 'all experiments have shown that electromagnetic and optical phenomena, relative to the earth as the body of reference, are not influenced by the translational velocity of the earth. The most important of these experiments are those of Michelson and Morley, which I shall assume are known. The validity of the principle of special relativity also with respect to electromagnetic and optical phenomena can therefore hardly be doubted' (quoted from [3]). Here it seemed that Einstein had wrongly separated the Mirror P and Mirror Q in the M-M Experiment as two different inertial systems for justifying the validity of 'the principle about inertial system' (i.e. 'the principle of relativity') that 'laws of physics hold for all inertial systems', which he thought of as a basis for the Postulate. Some even further thought that the Experiment had directly proved the Postulate, as light propagates along the two mutually perpendicular trips with equal velocity c. They all had failed to realise that the entire apparatus is one light source system.

In the next section, we will refresh the interpretation on the Doppler Effect to demonstrate that light indeed *simultaneously propagates in the overlapping non-light source systems with velocity different to c*, as described by the Emission Theory, and hence judge SR's Postulate is definitely wrong.

7. DOPPLER EFFECT
IS AN IRREFUTABLE
EVIDENCE DENYING SR

Let's first deduce the formulae of **Doppler Effect of sound** in order to contrast it with the Doppler Effect of light for better understanding.

Sound needs a medium as wave carrier. Vibration provided by a vibrator in the medium produces sound waves. But the true sound source is not the vibrator, because it cannot produce sound without a medium. Sound is born in the part of the medium where vibration occurs. That part of the medium is the real sound source. For example, a sound source is not the tuning fork (a vibrator) but is the part of the air surrounding it, because without air, the vibration of the fork cannot make sound. When a hammer hits a rail track, the part hit is the sound source that produces sound propagating through the other part of a rail track as a medium.

Distinguishing a sound source from a vibrator may help to understand that the sound waves' velocity in a medium could be different to the one with respect to the vibrator, or that there could be a relative velocity between the medium and the vibrator.

After being produced in the 'sound source part' of the medium, the sound continues to propagate in the medium. The magnitude of the propagating velocity c_s is completely determined by the property of the medium. For a certain medium, c_s ($= f \lambda$) is an unchangeable constant. While the frequency f of sound waves would be dependent on the vibration of the vibrator, the wavelength λ has to match the frequency f in forming the constant propagating velocity.

Thus the propagating velocity c_s of sound wave is defined to be relative to the medium and is independent of anything else including the motion of the vibrator (or a receiver). When the vibrator (or a receiver as well) has a relative velocity to the medium, the receiver would receive a sound wave with relative frequency different to the original one in the medium. That is referred to as the Doppler Effect of sound.

The velocities of the three: the vibrator, the sound wave, and the receiver, are all denoted with respect to the medium as v_r, c_s, and v_e, respectively. We shall only study the Doppler Effect in the situation when v_r and v_e are collinear.

We let c_s always being positive; let v_r being positive when it is pointing to the opponent of the approaching waves; and let v_e being positive when it is pointing to the receiver.

According to the concept of composite motion under classical mechanics, the velocity of the sound wave would be $(c_s - v_e)$ relative to the vibrator and would be $(c_s + v_r)$ relative to the receiver.

Our aim is to find out what frequency and wavelength of sound wave the receiver would receive.

CASE 1. $v_r = 0$ and $v_e \neq 0$

In this case, the receiver is at rest and the vibrator is moving relatively to the medium.

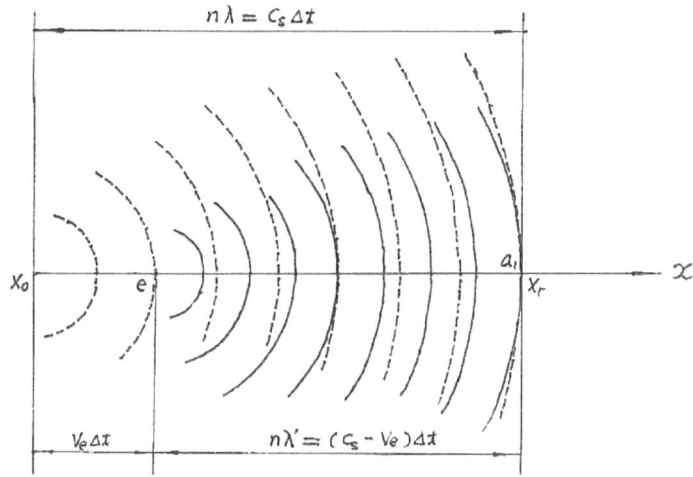

Fig.10 Sound's CASE 1 (in the stationary medium system)

At an arbitrary initial instant t_0 in the coordinate system of the stationary medium, let the Vibrator E be at point x_0 and mark an arbitrary spot x_r on the x-axis in the moving direction of the vibrator (Fig.10).

Suppose that the sound wave generated at t_0 by the Vibrator E from point x_0 reaches the arbitrary point x_r at instant t while in the same time $\Delta t = t - t_0$ the Vibrator E has moved from point x_0 to point e.

Obviously,

$$\mathbf{x_0 x_r} = c_s \, \Delta t \qquad \mathbf{x_0 e} = v_e \, \Delta t \qquad \mathbf{e x_r} = \mathbf{x_0 x_r} - \mathbf{x_0 e} = (c_s - v_e) \, \Delta t$$

From t_0 to t, the number of full waves generated with wavelength λ equals $n = \mathbf{x_0 x_r} / \lambda = (c_s \, \Delta t)/\lambda$. These waves would queue up on $\mathbf{x_0 x_r}$ if the vibrator does not move. But now, the moving vibrator brings the final Wavefront b to point e. Meanwhile, the first Wavefront a_1 by velocity c_s just arrives at point x_r. So, the same number of waves must be crowded into $\mathbf{e x_r}$ with wavelength changing from λ to $\lambda' = \mathbf{e x_r}/n = [(c_s - v_e)/c_s] \, \lambda$. This is because during the vibrator moves from x_0 to e, the stationary medium does not move. If a receiver is located at the arbitrary point x_r, these waves with compressed wavelength λ' are what the receiver encounters, and hence, the λ' equals the wavelength λ_{r1} the receiver would receive:

$$\lambda_{r1} = \lambda' = [(c_s - v_e)/c_s] \lambda \tag{26}$$

However, we need to demonstrate the physical compression process: how the same number of waves would be crowded into **ex**, with wavelength changing from λ to $\lambda' = [(c_s - v_e)/c_s] \lambda$. The demonstration starts from the first single wave (Fig.11):

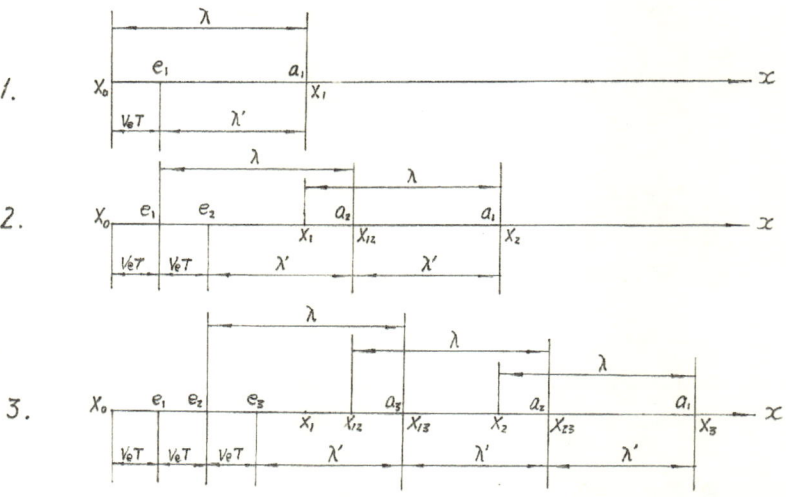

Fig.11 Sound's CASE 1 (the compression of wave length)

1. In the first time interval of one period T (= $1/f$ = λ/c_s), the first Wavefront a_1 emitted from origin point x_0 reaches at point x_1 by a displacement of one wavelength x_0x_1 = λ (= $c_s T$). Meanwhile, the Vibrator E arrives on point e_1 by a distance of x_0e_1 = $(v_e T)$ so that the e_1x_1 equals the new wavelength $\lambda' = \lambda - v_e T$;

2. In the second time interval of one period T (= λ/c_s), the second Wavefront a_2 emitted from vibrator point e_1 reaches at point x_{12} and the first Wavefront a_1 from point x_1 propagates to point x_2, both by the same displacement of one wavelength e_1x_{12} = x_1x_2 = λ. Meanwhile, the Vibrator E from point e_1 arrives on point e_2 by a distance of e_1e_2 = $(v_e T)$, so that the e_2x_{12} and the $x_{12}x_2$ both equals the new wavelength $\lambda' = \lambda - v_e T$;

3. In the third time interval of one period T (= λ/c_s), the third wavefront a_3 emitted from vibrator point e_2 reaches at point x_{13} and the second

and first wavefront: a_2 and a_1 from point x_{12} and x_2 propagates to point x_{23} and x_3, respectively, all by the same displacement of one wavelength $e_2x_{13} = x_{12}x_{23} = x_2x_3 = \lambda$. Meanwhile, the Vibrator E from point e_2 arrives on point e_3 by a distance of $e_2e_3 = (v_e T)$ so that the e_3x_{13}, the $x_{13}x_{23}$, and the $x_{23}x_3$ all equals the new wavelength λ' $= \lambda - v_e T$.

4. Continuing the process with one wave by another will produce a train of light waves with a compressed wavelength $\lambda' = \lambda - v_v T =$ $(c_s - v_e) T = [(c_s - v_e)/c_s] \lambda$.

The original wavelength λ indeed could be compressed into $\lambda' = [(c_s - v_e)/c_s]$ λ, as long as the vibrator has a velocity v_e relative to the medium.

It should be noted that it is the medium, as sound wave carrier, that makes the compression of wavelength possible. There would be no way to compress wavelength of light that has no medium to carry its waves.

As the receiver rests in the medium, the velocity of sound wave relative to the receiver must be equal to c_s. The train of the sound wave with compressed wavelength $\lambda_{r1} = \lambda'$ (26) must pass through the receiver with velocity of the c_s, and hence with frequency received by the receiver as

$$f_{r1} = c_s / \lambda_{r1} = [c_s /(c_s - v_e)] f \qquad\qquad (27)$$

We can see that the motion of the vibrator does not change the propagating velocity of sound wave in the medium in magnitude but has altered its wavelength and hence its frequency. The alteration of wavelength and frequency is due to the velocity of sound wave relative to the vibrator being changed from c to $(c_s - v_e)$.

A stationary receiver in the medium on the x-axis would encounter and receive the altered wave with $(f_{r1} \lambda_{r1}) = c_s$. This result may be referred to as a 'vibrator type' of Doppler Effect, since it is caused only by the motion of the vibrator in respect to the medium.

Notice that the results are independent of time and the relative position between the vibrator and the receiver. Thus, the result can be directly applied for any time and position.

CASE 2. $v_e = 0$ and $v_r \neq 0$

In this case, the vibrator is at rest and the receiver is moving relatively to the medium.

Think of a train of sound waves with wavelength λ approaching the moving Receiver R with velocity $(c_s + v_r)$ (Fig.12). The motion of the Receiver R does not influence the incoming waves' wavelength and hence the wavelength λ_{r2} the Receiver R would receive must be:

$$\lambda_{r2} = \lambda \qquad\qquad (28)$$

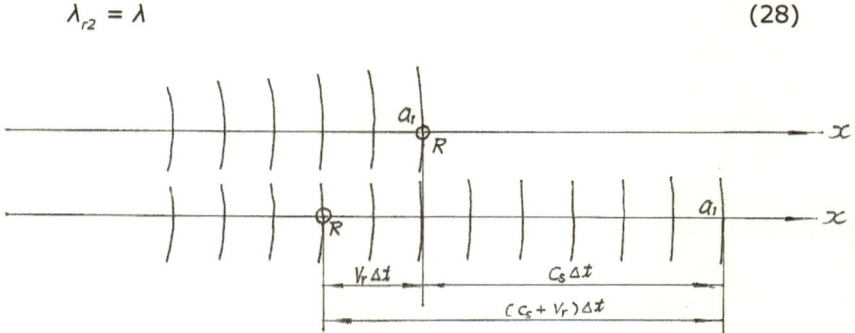

Fig.12 Sound's CASE 2 (in the stationary medium system)

The frequency of the sound waves in the medium also is not affected by the motion of the receiver. But due to the relativity of motion, the frequency with respect to the receiver would be different to the original one f in the medium. This is an apparent change in frequency only. In a time interval Δt, the train of waves passing through the receiver would have a total length of $(c_s + v_r)\Delta t$, and the number of these full waves equals $[(c_s + v_r)\Delta t / \lambda]$.

So, the apparent frequency received relatively by the receiver would be

$$f_{r2} = [(c_s + v_r)\Delta t / \lambda] /\Delta t = [(c_s + v_r) / c_s] f \qquad\qquad (29)$$

Check that $(f_{r2} \lambda_{r2}) = (c_s + v_r)$. This equals the relative velocity between the receiver and the wave. The result agrees with GT.

One can see that in this case, the moving receiver does not disturb the propagation of the waves in the medium but would relatively feel an apparent frequency different to the one in the medium (the apparent frequency would make no sense to a receiver of a non-sensor; and there would be nothing changing in the medium actually). Such a Doppler Effect is completely distinct to the one in Case 1. We may refer it as 'receiver type' of Doppler Effect, since it is caused only by the motion of the receiver in respect to the medium.

Notice that the results are independent of time and the relative position between the vibrator and the receiver. Thus, the result can be directly applied for any time and position.

CASE 3. $v_e \neq 0$ and $v_r \neq 0$

This is the general case that both the vibrator and the receiver are moving relatively to the medium. There would be a combinative Doppler Effect that is formed by combining Case 2 with Case 1.

From Case 1, we know that when the vibrator has motion relative to the medium, the disturbed wave would retain the same speed c_s but alter the wavelength from λ to λ_{r1} and hence change the frequency from f to f_{r1}, as expressed in the formula (26) and the formula (27) respectively.

From Case 2, we know that the moving receiver does not disturb the propagation of the waves on both wavelength and frequency and hence will accept the same wavelength of the incoming wave, i.e. $\lambda_{r2} = \lambda$ (28). But the moving receiver would relatively sense an apparent frequency f_{r2}, different to f, as expressed in the formula (29).

Since the results of the two cases are both independent of time and the relative position between the vibrator and the receiver, hence they can be composited for any time and position to fit in Case 3. All we should do is let the output sound waves from Case 1 be the incoming sound waves of the receiver and then handle it as in Case 2.

Therefore, the moving receiver will accept the wavelength λ_r, expressed similarly to λ_{r2} by Case 2's formula (28), but the wavelength of the incoming wave now is no longer the original wavelength λ. Instead, it should be substituted by the wavelength λ_{r1} of the wavelength-compressed wave, expressed in the formula (26) of Case 1. So the wavelength λ_r received by the moving receiver will be:

$$\lambda_r = \lambda_{r1} = [(c_s - v_e)/c_s] \lambda$$

The moving receiver will relatively sense an apparent frequency f_r, expressed similarly to f_{r2} by Case 2's formula (29), but the frequency of the incoming wave now is no longer the original frequency f. Instead, it should be substituted by the frequency f_{r1} of the wavelength-compressed wave, expressed in the formula (27) of Case 1. So the relative apparent frequency f_r sensed by the moving receiver will be

$$f_r = [(c_s + v_r)/c_s] f_{r1} = [(c_s + v_r)/c_s] [c_s/(c_s - v_e)] f$$

In this Case 3, we have derived the most general formulae for sound's Doppler Effect

$$\lambda_r = [(c_s - v_e)/c_s] \lambda \text{ and } f_r = [(c_s + v_r)/(c_s - v_e)] f \qquad (30)$$

Check that $(f_r \lambda_r) = (c_s + v_r)$. This equals the relative velocity between the receiver and the wave. The result agrees with GT.

One can see that v_e could not change the constant propagation speed but could alter both the wavelength and the frequency of the sound wave, which is the real part of frequency received by receiver; while v_r could not change the sound wave in the medium at all but would let the moving receiver hear the sound with an apparent part, in addition to that real part (generated by v_e), of the combinative frequency.

In all cases, the constant propagation speed of sound wave in the medium is always unchanged, because it is dependent upon the physical character of the medium only.

Note that in the case that both the receiver and the vibrator have velocity of equal magnitude relative to the medium but are stationary with respect to each other (i.e. $v_r = -v_e \neq 0$), the receiver would receive $\lambda_r = [(c_s - v_e)/c_s]\lambda$ and $f_r = f$. The combinative Doppler Effect still exists in reality, but just the particular receiver could not hear it (the apparent part of the combinative frequency has cancelled the real part).

Now we turn to investigate the **Doppler Effect of light**.

Since the propagation of light wave does not require a medium, there is no medium that could be referred to as the common reference system for the motion of the three: the emitter, the light wave, and the receiver. We have to take either the emitter or the receiver as the reference system.

As the emitter is a light source system, the light emitted from it must have the emitting velocity c relative to the emitter.

Let the velocity of light relative to the emitter always be positive and with its magnitude $c = f\lambda$ (where f is frequency and λ is wavelength).

Let the relative velocity v between the receiver and the emitter being positive when the two are moving close to each other.

Consideration 1. The emitter is taken as stationary reference system and the receiver moves toward the emitter with v.

The motion of the receiver does not disturb the incoming light's wavelength λ and hence the moving receiver R would accept the same wavelength

$$\lambda_r = \lambda \tag{31}$$

The apparent frequency f_r sensed by the moving receiver R can be derived as follows (Fig.13).

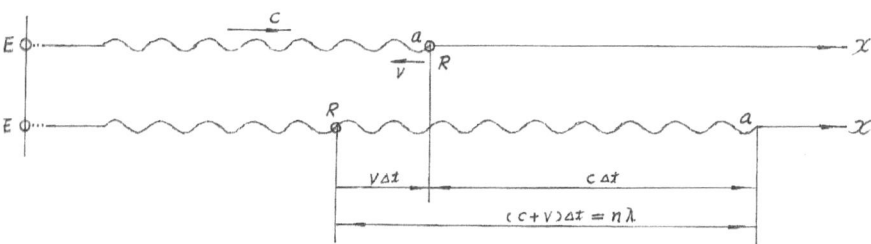

Fig.13 Light's Consideration 1 (in the system of the emitter **E**)

According to the concept of composite motion's velocity addition (under classical mechanics), the relative velocity between the light and the receiver should be $(c + v)$. In a time interval Δt, the total length of a train of light waves passing through the receiver would be $(c + v)\Delta t$ and the number of the full waves passing would be $n = [(c + v)\Delta t / \lambda]$. So, the receiver R would receive a relative frequency $f_r = n/\Delta t = [(c + v)\Delta t / \lambda]/\Delta t$, i.e.

$$f_r = [(c + v) / c] \, f \tag{32}$$

It indicates that, under classical mechanics, the apparent frequency f_r the moving receiver senses is greater than the incoming light's frequency f by a factor of $[(c + v)/c]$. Make no mistake. This is just a matter of relativity of motion. The propagation of the light wave relative to the emitter is not disturbed by the motion of the receiver at all. This is why this Doppler Effect is classified as 'receiver type', not 'emitter type' which would make disturbance on the wave.

If there is no relative motion between the receiver and the emitter (i.e. $v = 0$), there would be no Doppler Effect (i.e. $f_r = f$).

One can see that, light's *Consideration 1* is very much similar to sound's Case 2. The two situations both have no emitter's (or vibrator's) disturbance on the wave. In light's *Consideration 1*, the velocities of the light wave and the receiver are both relative to the same thing (the emitter), while in sound's Case 2, the velocities of the sound wave and the receiver are both relative to the same thing (the medium), too. As such, the two situations are both in the same way of composite motion's velocity addition to derive the relative velocity between the wave and the receiver. That is why their results are similar.

Consideration 2. The receiver is taken as stationary reference and the emitter moves toward the receiver with v.

In the coordinate system of the stationary receiver (Fig.14), suppose that at an initial instant, a Wavefront a_1 just arrives on the receiver R at an arbitrary spot located in the moving direction of the emitter E that is carrying a light source system as its attached space (including the whole electromagnetic field), in which light propagates with the emitting velocity c. In an arbitrary time increment Δt, the moving emitter carries the whole attached space forwards in the stationary reference system of the receiver passing through the receiver R by a distance of $(v \, \Delta t)$. Meanwhile, the Wavefront a_1 would propagate forward in the emitter's attached space by a distance of $(c \, \Delta t)$. Total length of the train of waves passing through the receiver R equals $(c +$

v) Δt (under the concept of composite motion). One can see that there is no compression of wavelength occurring and hence no 'emitter type' of Doppler Effect happening.

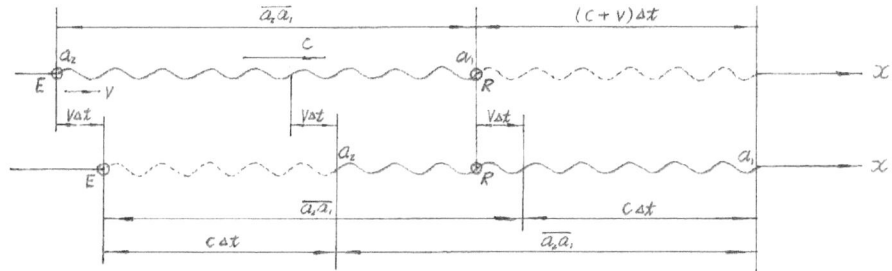

Fig.14 Light's Consideration 2 (in the system of the receiver **R**)

Light's *Consideration 2* seems to be similar to sound's Case 1 but actually turns out contrary to it. Sound can have 'vibrator type' Doppler Effect, but light can't ('emitter type'); sound has no 'receiver type' Doppler Effect in Case 1, but light does in *Consideration 2*.

Why could sound have a 'vibrator type' of Doppler Effect but light could not? A sound wave is a longitudinal wave and must be carried by the medium that is separate from the vibrator. These provide conditions for wave compression. When the sound receiver is resting in the medium (Case 1), the moving vibrator, together with part of the medium as a sound source, can move forward to catch up with the newly emitted wavefront such that the wavelength reduces and the wave is compressed (Fig.10). So, the 'vibrator type' of Doppler Effect would occur for sound. But for light waves, as transverse non-substance waves with no need for a wave carrier (medium), it is absolutely impossible to get its electromagnetic field compressed by the effect of mechanics. In theory, the emitter is unable to interfere with its own attached space, and hence, there is no way for the moving emitter to make any influence on the light waves and their electromagnetic field. Therefore, the emitter's motion is useless in generating 'emitter type' of Doppler Effect for light.

Nevertheless, they do have a 'receiver's type' of Doppler Effect, since a train of light waves have passed through the receiver R at an arbitrary point. The situation is exactly the same as in *Consideration 1*. Since the train of light waves retain the original wavelength λ without compression, the receiver would accept the unchanged wavelength (31) but an apparent frequency (32) under the concept of composite motion (classical mechanics):

$$\lambda_r = \lambda \quad \text{and} \quad f_r = [(c + v) / c] \, f \tag{33}$$

It is no wonder that the *Consideration 1* and *Consideration 2* have the same result, because they just swap the status of relative motion between the emitter and the receiver.

The result indicates that light can only have a 'receiver's type' of Doppler Effect as shown in the (33) but no 'emitter type' (Note that 'emitter type' of Doppler Effect is specifically defined as the one caused by the disturbance of the emitter on the emitted light), whether we take the *Consideration 1* or the *Consideration 2* (under classical mechanics).

The 'receiver's type' of Doppler Effect for light depends on the receiver's velocity relative to the emitter, while for sound it depends on the receiver's velocity relative to the medium. Is there any common point for both? Yes. The emitter for light is the light source, while the medium for sound has the 'sound source part' surrounding the vibrator as the true sound source and, hence, the receiver's velocity relative to the medium also is relative to sound source. So, for both light and sound, the receiver's velocity can be defined as relative to the source. Therefore, whether for light or for sound, the 'receiver's type' of Doppler Effect has no difference.

Note that all results above are under classical mechanics and need not to employ GT. Now we turn to consider using LT under SR.

Take the emitter as the stationary light source Galilean System A and let the emitter be seated at the origin A and emit light starting at a zero initial instant. The Receiver R is moving with *v* relative to the emitter along the positive direction of the *x*-axis and has an arbitrary distance from the emitter when at the initial instant. Let the Receiver R carry its attached space as the moving Lorentz System B. It doesn't matter what location ξ_R the Receiver R, as the second party, is stuck on the ξ–axis in System B, as long as the origin B is coinciding with origin A at the initial instant for the standard configuration. We may find out what frequency and wavelength the light in the moving System B would be, by applying LT on the light (Fig.15).

Suppose that, at an arbitrary instant t, the whole System B has moved a displacement of (vt) while the initial flash is forming a wavefront sphere of radius r in System A. The number of full waves along any radial Ray AP would be $n = r/\lambda = c\,t/\lambda = f\,t$ (where λ and f is the wavelength and frequency of the light in Galilean System A). Correspondingly, in Lorentz System B, Ray BP should have the same number of full waves and hence would have wavelength and frequency:

$$\lambda' = \rho/n = c\,\tau_p/(c\,t/\lambda) = \lambda\,[\beta(t - x_p\,v/c^2)/t] = \lambda\,\beta[(c - v\cos\varphi_x)/c]$$

$$f' = n/\tau = f\,t/\beta(t - x_p\,v/c^2) = f/\beta[(c - v\cos\varphi_x)/c] = f\,c/\beta(c - v\cos\varphi_x)$$

where $x_p = c\,t\cos\varphi_x$ and φ_x is the direction angle of the Ray AP over the x-axis.

Check that $f'\lambda' = f\lambda = c$, as required by SR's Postulate. The results demonstrate both the wavelength and the frequency of light in the moving System B are different to those in stationary System A and are also different from ray to ray, as φ_x would change. Such result reflects the strange features of Lorentz system.

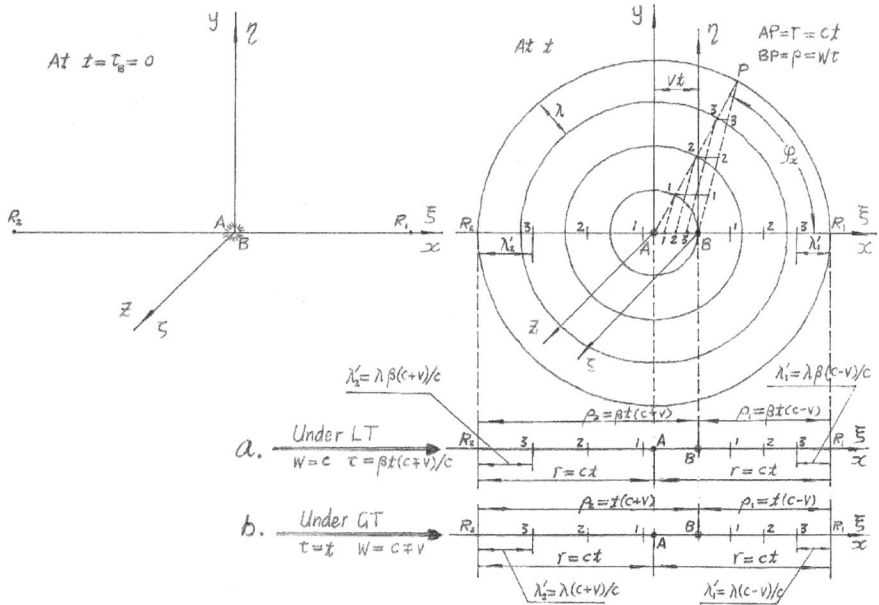

Fig.15 (a) 'LT effect' (b) 'GT effect'

Suppose that, at an arbitrary instant t, the whole System B has moved a displacement of (vt) while the initial flash is forming a wavefront sphere of radius r in System A. The number of full waves along any radial Ray AP would be $n = r/\lambda = c\,t/\lambda = f\,t$ (where λ and f is the wavelength and frequency of the light in Galilean System A). Correspondingly, in Lorentz System B, Ray BP should have the same number of full waves and hence would have wavelength and frequency:

$$\lambda' = \rho/n = c\,\tau_p/(c\,t/\lambda) = \lambda\,[\beta(t - x_p v/c^2)/t] = \lambda\,\beta[(c - v\cos\varphi_x)/c]$$

$$f' = n/\tau = f\,t/\beta(t - x_p v/c^2) = f/\beta[(c - v\cos\varphi_x)/c] = f\,c/\beta(c - v\cos\varphi_x)$$

where $x_p = c\,t\cos\varphi_x$ and φ_x is the direction angle of the Ray AP over the x-axis.

Check that $f'\lambda' = f\lambda = c$, as required by SR's Postulate. The results demonstrate both the wavelength and the frequency of light in the moving System B are different to those in stationary System A and are also different from ray to ray, as φ_x would change. Such result reflects the strange features of Lorentz system.

Particularly for the rays lying on the x-axis with $\varphi_x = 0°$ and $180°$ in System A, their corresponding rays that lie on the ξ-axis in System B would have wavelength λ' and frequency f' as below (Fig.15(a)):

- For $\varphi_x = 0°$

$$\lambda' = \lambda\,\beta(c - v)/c = \lambda\,\sqrt{(c-v)/(c+v)} \qquad\qquad (34\ a)$$

$$f' = f\,c/\beta(c - v) = f\,\sqrt{(c+v)/(c-v)} \qquad\qquad (34\ b)$$

- For $\varphi_x = 180°$

$$\lambda' = \lambda\,\beta(c + v)/c = \lambda\,\sqrt{(c+v)/(c-v)} \qquad\qquad (34\ c)$$

$$f' = f\,c/\beta(c + v) = f\,\sqrt{(c-v)/(c+v)} \qquad\qquad (34\ d)$$

Notice that the λ' and f' are independent of the time and the distance between the receiver and the light source. Hence, they are the wavelength and frequency of light travelling along the whole ξ-axis in System B.

It has been said that the λ' and f' in (34 a, b, c, d) are the Doppler Effect (under SR) received by the moving receivers: R_1 and R_2 (on the right side of origin A is R_1 recessing from A while on the left side is R_2 approaching toward A). Wrong, they are not. They are just a kind of 'LT effect' that is a **real** change in wavelength and frequency on every direction happening on the light in Lorentz System B under LT. This 'LT effect' is a matter of light's motion transformation reflecting the strange features of Lorentz space and Lorentz time. It does not address the physical meaning of Doppler Effect ('receiver type').

Distinctively in substance, Doppler Effect ('receiver type') is sensed by a moving receiver on an **apparent** change in frequency of the incoming light in Galilean System A. The original wavelength of light would not be disturbed by the motion of the receiver.

Make no mistake. The receivers are the second party bodies carrying their common System B. As we have explained in Section 1, second party possesses the space and time of first party System A only and cannot experience the space and time of its own attached System B, i.e. cannot experience the λ' and f' of the (34 a and b or c and d). The light is travelling along AP with $u = c$ in Galilean System A where light has λ and f, while simultaneously along BP with $w = c$ in Lorentz System B where light has λ' and f'. The second party receivers with v would encounter the light with $u = c = \lambda f$ in System A, but definitely could not meet the light with $w = c = \lambda' f'$ in System B.

Therefore the 'LT effect' is certainly not the Doppler Effect. In other words, Doppler Effect does not need to employ LT at all. The difference is that, the 'LT effect' focuses on the motion transformation of the light as a third party between the two systems of emitter and receiver, while the Doppler Effect focuses on the second party receiver's sense on the incoming light as another second party with respect to their common reference of first party emitter.

Even in classical mechanics, Doppler Effect also does not need to employ GT. Following is the explanation.

If try to apply GT on the light between the two systems as shown in Fig.15 (b), we have $\tau = t$ and light's relative velocity $w = c \mp v$ (for ray BR_1 or BR_2 on the ξ-axis in System B). The number of waves queuing up on the ray AP in System A is $n = r/\lambda = c\,t/\lambda$. Since the Ray BP in System B must correspondingly have the same number (n) of waves queuing up on it, the wavelength and frequency of ray BR_1 or BR_2 on the ξ-axis in System B would be

$$\lambda' = p/n = w\tau = (c \mp v)\,t\,/(c\,t/\lambda) = [(c \mp v)/c]\,\lambda$$

$$f' = n/t = f$$

However these are not like the Doppler Effect because Doppler Effect must alter the frequency of the incoming light. Actually these are the 'GT effect' on the light's wavelength (the effect from light's velocity transformation under GT) reflecting the influence of the v of the moving System B. The receivers, as second party, are unable to experience the 'GT effect' that is for third party exclusively.

The true classical mechanics Doppler Effect is the (33): $\lambda_r = \lambda$ and $f_r = [(c \mp v)/c]\,f$ (the minus sign is for the receiver recessing), which is formed on the basis of composite motion's velocity addition in System A for the receivers.

We can see that, under classical mechanics, 'GT effect' and Doppler Effect conform to each other on velocity as $w = \lambda' f' = \lambda_r f_r$. Nevertheless, 'GT effect' does not change light's frequency but alter light's wavelength, while Doppler Effect does not change light's wavelength but alter light's frequency.

It is clear that neither the 'LT effect' nor the 'GT effect' is Doppler Effect which is in substance a matter of composite motion's velocity addition and needs not to employ any kinematic transformation, neither LT nor GT, at all.

One can see that misapprehending 'GT effect' or 'LT effect' as Doppler Effect is due to mistaking second party as third party. For SR, it is the same conceptual error of 'applying LT wrongly on second party'.

Now let's forget the 'LT effect' but continue to find any possible Doppler Effect under SR's Postulate.

We have previously explained why light could not have 'emitter type' of Doppler Effect by virtue of the concept of a light source system, which is accepted by both the classical mechanics and the SR's Postulate. An emitter can never disturb the propagation of light waves by compressing its wavelength. This is because light always retains the emitting velocity c relative to the emitter, which is right the meaning of Einstein's second principle of 'the constancy of the velocity of light' (see Section 8). Therefore, not only the classical mechanics, but also SR, would agree that there is no 'emitter type' of Doppler Effect for light (Note that 'emitter type' of Doppler Effect is specifically defined as the one caused by the disturbance of the emitter on the emitted light). Should any Doppler Effect occur for light, it must be 'receiver type'.

To investigate the 'receiver type' Doppler Effect for light under SR, we can refer to the previous result under classical mechanics: $\lambda_r = \lambda$ and $f_r = [(c + v) / c] f$ (33) with proper modification to suit SR's Postulate.

Both the receiver and its incoming light are moving in the emitter's Galilean System A, where they encounter each other. Only there, the light can be sensed by the receiver. As light is an electromagnetic wave, it would not be disturbed by the mechanical motion of the receiver; and when it is passing through, the receiver has to accept its wavelength fully in Galilean space of System A, i.e. $\lambda_r = \lambda$.

Since the receiver is moving with v relative to the light source (emitter), the light's velocity relative to the receiver would be $(c + v)$ if under classical mechanics, and hence the receiver should 'feel' a changed f_r as we have already derived in the (30).

Nevertheless, SR's Postulate stipulates that light always has c relative to everything. Consequently, the relative velocity between the light and the receiver now has to be taken as c, instead of $(c + v)$. Here it is the Postulate taking over the concept of composite motion of classical mechanics. In a time interval Δt, under SR, the total length of a train of light waves passing through the receiver would be $(c\Delta t)$, and the number of the full waves passing would be $(c\Delta t / \lambda)$. So, the receiver would receive a frequency $f_r = (c\Delta t / \lambda) / \Delta t = f$.

The result of $\lambda_r = \lambda$ and $f_r = f$ means that there is no 'receiver type' of Doppler Effect for light under SR's Postulate.

It is shocking but understandable. Considering that light always has the same velocity of c relative to both the light source and the receiver (under SR), the two are on completely an equal footing. Since there must be no 'emitter type' Doppler Effect for light, of course no 'receiver type' would occur either. In reality, Doppler Effect would be generated only when light's propagating velocity V relative to the receiver is different to light's emitting velocity c (that is the situation under classical mechanics). When the V must equal c, as stipulated by the Postulate, the situation is just like you are standing firmly on the ground and receiving light from a lighthouse with no Doppler Effect occurring.

Therefore, if SR's Postulate is right, light's velocity relative to the receiver must always be c, the light the receiver would sense must be as normal as the original light; then no Doppler Effect could happen. In other words, *SR's Postulate and Doppler Effect are bound to be mutually exclusionary in nature*. Any calculation for Doppler Effect using SR (Einstein had done it in *§7. Theory of Doppler's Principle and of Aberration* of [1]) surely makes no sense at all (It is a good example signifying that mathematics calculations must be subject to physics conditions first).

It really is startling. The fact that there certainly is Doppler Effect for light found in astronomy is irrefutable evidence denying the Postulate. Hence, SR must be wrong and Doppler Effect for light being 'receiver type' can only be shown in (33) under classical mechanics:

$$\lambda_r = \lambda \quad \text{and} \quad f_r = [(c + v) / c]\, f \tag{33}$$

Check that $f_r \lambda_r = (c + v)$, which indicates that the light's velocity, relative to the receiver, is not the 'absolute c' but is 'c plus'!

It clearly indicates that light always propagates with emitting velocity c only in the field of the light source's attached space, no matter what velocity the light source would have relative to anything else. Meanwhile, the light *simultaneously propagates in the overlapping non-light source systems with velocity different to c*, which verifies that the light's Emission Theory and classical mechanics are correct while the Postulate and SR are wrong.

Also, we can see that c is generally not a limiting value for relative velocity. Anybody can have a velocity greater than c, relative to light, as long as it moves in the direction opposite to the light. In general, classical mechanics' velocity addition formula for composite motion gives no limit to the relative velocity.

However, under the influence of electrical field, so far no charged particle can gain a velocity exceeding c. But that is a different story. Here, the velocity of a particle is relative to the field exclusively. If anybody moves in the direction opposite to the fast moving particle, the relative velocity between the body and the particle could exceed c. As to why would the charged particle's velocity (relative to the field) have a limiting value equal to c, it is yet to be explained. Perhaps the resulting force exerted on the particle is velocity-dependent, for there may be an anti-effect caused by the motion of the particle to repel the local field: As the velocity of the particle becomes greater, the anti-effect grows stronger; when the particle's velocity is close to c, the anti-effect grows rapidly so that the resulting force exerted on the particle eventually approaches zero and hence no more acceleration in the limit.

8. THE POSTULATE WAS BORN WITHOUT REASONABLE LOGIC

So far, we have discovered that LT has many defects, including the dependency on light source, the uncertainty by configuration, the two deadlocks in space and time, the contradiction between magnitude and direction of relativistic vectors, the incompatibility with Newton' laws, and the false local-time of no flux. We also have found that the SR's Assertion of 'moving rod shortens, moving clock slows' or the relativistic mechanics was improperly made with a serious conceptual error of 'applying LT wrongly on second party'. All this shows that SR is untenable in theory.

In analysing the M-M Experiment, we have clarified there was no support for SR. In reinterpreting Doppler Effect, we have revealed that the Doppler Effect and the Postulate are bound to be mutually exclusionary in nature. The existence of Doppler Effect found in astronomy convincingly indicates that the Postulate is definitely unrealistic.

Clearly, our conclusion must be that, with so many fatal problems, SR is a pseudoscience.

For a long period of time, there has been a prejudice that LT is a general law while GT is only an approximation for $v \ll c$. Now the verdict should be reversed. The truth is that the higher the v, the more serious the deviation of LT. Look back at Fig.4, if $v \to c$ in the limit, the origin B would end up coinciding with the point of the wavefront on x-ξ axis, as such, under LT, the two coinciding points would have velocity of c relative to each other. How ridiculous it is!

As we have pointed out in the beginning of the text, SR's Postulate meant that 'a light always has an absolute velocity c relative to all inertial systems simultaneously'; this Postulate is so vital that it plays a role of the pillar of SR. But it seemed also so strange that there was no sight of such an important statement in Einstein's formal paper works.

Even in his first SR paper (1905) [1], Einstein did not directly write his Postulate in words. Instead, he only provided two 'principles'.

He called the first one the 'principle of relativity' that 'the same laws of electrodynamics and optics will be valid for all coordinate systems in which the equations of mechanics hold' [1]. The first principle was also called the 'principle of special relativity' or the 'principle of relativity of translation', stating 'the laws of nature are in concordance for all inertial system' (quoted from [3]). We would like to rename it as the *'Principle about Inertial Systems'* in order to avoid confusing it with others named with relativity.

Note that this first principle does not govern the kinematic transformation between systems. A third party body has a velocity $\boldsymbol{u}$ in one system, and would have a different velocity $\boldsymbol{w}$ in another system. The two velocities would not be the same. Ball tossing vertically in a train would be seen as projectile motion from the ground. For the light's emitting velocity c in the light source system, the first principle does not mean this light would also have the c in another system. But Einstein attempted to make use of the first principle in such a way for his Postulate. Actually, the first principle could mean that every light source in its own system always emits light with the same emitting velocity c (That was just what Maxwell's equation told after the dissolution of ether).

Einstein named his second principle the 'principle of the constancy of the velocity of light' that 'Light always propagates in **empty space** with a definite velocity V that is independent of the state of motion of the emitting body' (quoted from [1]).

The statement of the second principle is rather confusing. It could *at least* mean 'light always propagates in **light source system** with a definite velocity $V = c$ that is independent of the state of motion of the emitting body', or directly, 'light's propagating velocity V in light source system is the same as the emitting velocity *c'*, which is exactly in conformity with the concept of light source system (i.e. the first part of the light's Emission Theory of classic mechanics). But, instead of the light source system, the term 'empty space' was used. Since everything is contained in 'empty space' and is touching the 'empty space and is resting with the 'empty space' where the all-pervading ether just left, you may further take *a revolutionary step* approaching to that 'light always propagates in *empty space* with a definite velocity *V = c* relative to everything', which has arrived on the same meaning as the Postulate. This may be the reason why his second principle was named 'the constancy of the velocity of light'.

What was the true mind of Einstein in his second principle?

'On 25 April 1912, in a letter to Paul Ehrenfest commenting on Ritz's emission theory, Einstein referred to 'Ritz's conception, which before the theory of relativity was also mine. He expanded on this remark on 20 June: "I knew that the principle of the constancy of the velocity of light was something quite independent of the relativity postulate, and I weighed which was more probable, the principle of the constancy of c, as required by Maxwell's equations, or the constancy of c exclusively for an observer located at the light source. I decided in favour of the former" (quoted from the editor's words in [1]).

Here, the former links with the idea of 'empty space', while the latter links with the idea of 'light source system'. As the former includes the latter **at least**, Einstein "decided in favour of the former". That indicated his Postulate was developed from the idea of 'empty space' while 'light source system' was excluded in his mind.

In the [3] 1922, Einstein clarified his logical thinking formally as follows: 'The consequence of the Maxwell-Lorentz equations that in a vacuum, light is propagated with the velocity c, **at least** with respect to **a definite inertial system K** (that obviously is the missing light source system), must therefore be regarded as proved (his second principle). According to the principle of special relativity (his first principle), we must also assume the truth of this principle (his second principle) for **every other** inertial system (approaching to the Postulate).'

Einstein still did not directly draw the revolutionary Postulate in words. We do not know why. But we do know there was an obvious problem in logic. The situation of the second principle that 'in a vacuum, light is propagated with the velocity c, **at least** with respect to **a definite inertial system K**' cannot be expanded by the first principle to '**every other** inertial system' such that '*this* light also propagates with $V = c$ relatively to all other systems simultaneously'. This is because the first principle can expand among inertial systems for dynamics laws only but not for kinematics quantity like velocity. In substance, the c is just the constant value of light's emitting velocity determined by Maxwell's equations in light source system only (after the dissolution of ether), while light's propagating velocity V in any system is depended on the relative motion between that system and the light source system. Einstein seemed mixing the concept of emitting velocity c with the concept of propagating velocity V.

We can see that the logic for SR's Postulate was broken in first. The Postulate had no chance to be inferred by the combination of his two principles. No wonder SR would end up to be a pseudoscience, as it started from the Postulate of no underlying substratum.

We also can see that Einstein just did not like the concept of light source system which would destroy the purity of the united world of Lorentz systems and consequently destroy SR's reliability. That was why the light source system was intentionally excluded from SR. Einstein never formally mentioned the term of 'light source system' in SR. Instead, he preferred to call it 'a definite inertial system k' in his direct path to SR (as just stated above) and then kept it apart from SR. But in practice, he had no choice but to always endow the stationary system with systemic time t and 'even' space $[x, y, z]$, the same as a light source Galilean system; otherwise, the interlocking rings on space and time between Lorentz systems could not be unlocked. However, he never mentioned that his stationary system actually was a Galilean system of the light source. It really was strange that a theory did not want to face the truth.

CONCLUSION

In this text, we have shown that:

- The concept of overlapping attached space and the identity of universe instant are two most important primary concepts of space and time, forming the foundation of assessment on SR.

- The missing light source Galilean system is the stepping stone for exploring the space and time in Lorentz system. Without it, we would still be trapped in the maze of LT. But the dependency on it makes LT uncertain and untrustworthy.

- The (15): $\tau = t/\beta - \dfrac{V}{C^z} \xi$ has been neglected by SR, but it is the most vital relation for dissecting Lorentz time properly.

- The (14): $\Delta\xi = \beta \Delta x$ (when $t = constant$) and the (16): $\Delta\tau = \Delta t/\beta$ (when $\xi = constant$) are the two fundamental relations for analysing Lorentz space and systemic time in contrast with Galilean system. No way for SR to resolve these two deadlocks.

- The deluding local-time part of Lorentz time, ($-\dfrac{V}{C^z} \xi$), is the most fatal fallacy of LT. Once it is pulled out, the whole building of SR will collapse immediately.

- The most serious conceptual error of 'applying LT wrongly on second party body' dominated the entire SR, including the relativistic mechanics, creating a fantastic world full of paradoxes, which are interesting but unrealistic.

- The M-M Experiment not only denies ether but also verifies the concept of light source system. While many scientists had puzzled at its negative result, it had nothing to do with LT and SR in reality.

- The Postulate and Doppler Effect are bound to be mutually exclusionary in nature, and hence the Doppler Effect found in astronomy provides the convincing evidence denying the Postulate definitely.

- The Postulate was born without reasonable logic in the first place. No wonder SR turns out to be a pseudoscience.

It is the time for science to wake up now. Nature does not listen to big talks. Mathematics tools cannot take over physical concepts. No Postulate, no LT and no SR for sure. What is next? As no Minkowski's space-time continuum in objective reality, it may end up with no Einstein's General Theory of Relativity eventually. All happening are just because the sky-high twin tower of the theories was built on a foundation of 'empty space' (left by ether)!

For a long period of time, science had been hazed by the concept of ether, until it was ascertained that the concept was false. The dissolution of ether would have given a chance for physics to go back on the right track with the concept of light source system and GT. But history had ever made jokes. Lorentz was reluctant to see the departure of ether. He invented LT under his Ether Theory with the hope that the M-M Experiment could be well explained with 'Lorentz Contraction' in the presence of ether, but the rescue of ether was eventually in vain. Einstein picked up the 'empty space' left by ether. Instead of ether, he let light propagating with c in 'empty space' which could be linked with everything. From the term of 'empty space' he set the Postulate as SR's pillar. He also took up the equations of LT to support the Postulate in a unique and strange way, and then somehow made the selling of SR very successful. As such, physics did not go back on the right track; rather, it went even further away from the truth. It is hard to believe that the term 'empty space' can follow after the ghost of ether continually haunting our poor physics for over a century.

However, the time has come to end the long, long drama of 'The Emperor's New Clothes' ultimately.

BIBLIOGRAPHY

[1]. 'On the Electrodynamics of Moving Bodies', Albert Einstein (1905). Collected in the book *Einstein's Miraculous Year-Five Papers that Changed the Face of Physics*, Princeton University Press, New Jersey, 1998; The Softback Preview, 1999.

[2]. *Relativity: The Special and General Theory,* Albert Einstein (1916). Translated by Robert W. Lawson, Dover Publications, New York, 2001.

[3]. *The Meaning of Relativity,* Albert Einstein (1921). Princeton University Press, New Jersey, 2005.

[4]. *Einstein's Theory of Relativity,* Max Born (1924). Dover Publications, New York, 1962.

[5]. *Introduction to the Theory of Relativity,* Peter Gabriel Bergmann (1942). Dover Publications, New York, 1976.

[6]. *Elements of Relativity Theory,* D. F. Lawden (1985), Dover Publications, New York, 2004.

INDEX

V

Z

www.ingramcontent.com/pod-product-compliance
Lightning Source LLC
Chambersburg PA
CBHW022115170526
45157CB00004B/1654